Combined Proceedings International Plant Propagators' Society 2020

IPPS Volume 70

Held 2020

Editors:

Robert L. Geneve
Clive Larkman
Charles W. Heuser, Jr.
Spencer Gunn
Masanori Tomita
Hayden Foulds
Elsa du Toit
Fred T. Davies, Jr.
Ryan Stewart

ISBN: 978-1-7138-3450-2

Printed from e-media with permission by:

Curran Associates, Inc.
57 Morehouse Lane
Red Hook, NY 12571

Some format issues inherent in the e-media version may also appear in this print version.

Printed with permission by Curran Associates, Inc. (2021)

For permission requests, please contact International Plant Propagators' Society (IPPS) at the address below.

International Plant Propagators' Society (IPPS)
174 Crestview Drive
Bellefonte, PA 16823-8516
USA

Phone: +01 814-355-2467

secretary@ipps.org

Additional copies of this publication are available from:

Curran Associates, Inc.
57 Morehouse Lane
Red Hook, NY 12571 USA
Phone: 845-758-0400
Fax: 845-758-2633
Email: curran@proceedings.com
Web: www.proceedings.com

THE INTERNATIONAL PLANT PROPAGATORS' SOCIETY

Combined Proceedings

Volume 70

2020

IPPS

Sharing Plant Production Knowledge Globally

Published 2020

Combined Proceedings International Plant Propagators' Society (ISSN 0538-9143) is a yearly publication of the International Plant Propagators' Society (IPPS) headquartered in Pennsylvania, U.S.A.
Email: Secretary@ipps.org
Website: www.ipps.org

Business Office. Memberships are processed at the regional level, with dues being allocated to both international and regional expenses. Contact the regional secretary / treasurer in your area of the world for membership information and applications.

Table of Contents

Southern Africa Region

Southern Region - USA

Western Region - USA

Combined Proceedings IPPS 70:1-5. 2020.

Quest for the Grassland Jewels

M. S. Deutschlander and Q. Bersiks

Department of Environmental Sciences; Ornamental Horticulture; Unisa, Florida Campus, C/O Christiaan de Wet and Pioneer Avenue, Florida, 1710.

deutsms@unisa.ac.za

Keywords: Highveld Grassland; Ornamental Horticulture species; Eco-friendly landscaping species; Medicinal plant species; propagation; cultivation

Summary

The Ornamental Horticulture programme group, Department of Environmental Sciences, UNISA, identified the need to recognize plant species of the Highveld region with ornamental, medicinal and eco-friendly landscape value for propagation and cultivation. According to the World Wild life fund 2020 the status of the Highveld grasslands is critical endangered due to severe agricultural fragmentation and urbanisation. These factors as well as climate change and extreme weather phenomena necessitate to be proactive in the identification and propagation of plant species best suited and resilient to extreme climate conditions. Not only to full fil the demands and needs of the green industry but also for the preservation of biodiversity. Many of these resilient characteristics already exhibited by many plant species of the grassland biome.

By making use of structured questionnaires and pilot studies, public inputs will be obtained to determine which plants are selected and preferred by the public and why. Results obtained will provide guidelines of which plants to consider for propagation and cultivation. The distribution and availability of selected plant material for propagation will also be investigated.

This is an umbrella project consisting of several individual MSc and PhD research projects investigating the propagation and cultivation of grassland plant species. Results will be compiled into and extensive volume on the propagation of Highveld grassland plant species.

IPPS Vol. 70 – 2020

INTRODUCTION

South Africa is the third most biodiverse country in the world, containing 10% of the worlds' floral biodiversity. Consisting of approximately 23 000 plant species, attributed to both a diverse climate and landscape (Koekemoer *et al.*, 2013, Van Wyk & Gericke, 2000). The grassland biome is the second largest biome in South Africa comprising 24,27 % of South Africa's surface (Palmer & Ainslie, 2005). The Highveld region contains more or less 3 370 plant species (Bond, 2016) of which only 250 are known to be cultivated commercially for ornamental and landscape use (Wentzel, 2019).

Many indigenous plant species found in the grassland biome are adapted and resilient to extreme weather conditions experienced in this summer rainfall area that encounters major changes in rainfall (400 – 900mm) and seasonal temperature (-11°C winter to 38°C summer) as well as large differences in daily temperatures (3-6°C - 21-24°C) (Bowie & Frank, 2020; Mucina & Rutherford 2006). Making them excellent plant specimens for eco-friendly gardening and landscaping, especially in Gauteng. The Highveld Grassland has suffered extensive degradation due to agricultural fragmentation as it is one of the best areas for farming practises (Bowie & Frank, 2020). It is also the most densely populated biome in South Africa containing, Gauteng Province, the economic hub of South Africa. Expanding human populations, urbanisation, diminishing natural habitats, over exploitation and unsustainable harvesting practises threatens the survival of many grassland plant species as well as the occurrence of invasive plant species competing with indigenous species for habitat and resources (Bowie & Frank, 2020).

The identification of grassland species with ornamental, medicinal and eco-friendly landscaping values need to be exploited to contribute to the survival of the species and maintenance of bio- and genetic diversity. Contributing to the economic growth of South Africa in creating job opportunities, export possibilities and research prosperities.

AIM AND OBJECTIVES

The aim of this project is to establish an Ornamental Horticulture research Group in the Department of Environmental Sciences at the University of South Africa for the propagation of indigenous plants occurring in the grassland biome with ornamental horticulture, medicinal and/or eco-friendly landscaping value.

Objectives of this project is to identify indigenous plant species with suitable characteristics and traits for ornamental horticultural and eco-friendly landscaping use as well as with medicinal value in the Highveld grassland. The investigation of optimal propagation methods and conditions for the cultivation of these identified plant species *via* cuttings, seeds or tissue culture. Ensure the survival of these species preserving biodiversity and possible commercialisation.

LITERATURE REVIEW

An extensive literature review was done on plants that occur in the grassland biome of South Africa, specifically the Gauteng region, comprising the Soweto Highveld Grassland (GM 8), Egoli Granite Grassland (GM 10) and the Rand Highveld Grassland (GM 11) (Mucina & Rutherford, 2006). Indigenous plant species characteristic of the grassland biome that exhibit desired traits for ornamental horticulture and eco-friendly landscaping was identified, being it structure, foliage or flower production, growth form or various other factors.

By making use of structured questionnaires, public inputs were obtained concerning plants selected by the public. Provided an indication of plants to consider for propagation.

Rotheca hirsuta (Hochst.) R.Fern.
Family: Lamiaceae
Clematopsis scabiosifolia (DC.) Hutch.
Family: Ranunculaceae
Pachycarpus schinzianus (Schltr.) N.E.Br.
Family: Apocynaceae
Macledium zeyheri (Sond.) S.Ortiz
Family: Asteraceae
Dicoma anomala Sond. Family: Asteraceae
Pentanisia prunelloides (Klotzsch ex Eckl. & Zeyh.) Walp. Family: Rubiaceae
Boophone disticha (L.f.) Herb
Family: Amaryllidaceae
Chironia palustris Burch. subsp. Palustris
Family: *Gentianaceae*
Lannea edulis (Sond.) Engl. var. *edulis*
Family: Anacardiaceae

METHODOLOGY

Three grassland plant species were selected from the above initial plant list compiled for this pilot study to determine the public opinion. The pilot study was conducted by making use of questionnaires to determine the publics opinion on which of the three-plant species selected and presented, they found the most attractive and would want to plant in their gardens and the reasons why.

Surveys

Surveys were done at various garden centres in the Johannesburg area and the SANA trade day by making use of face to face surveys. An e-mail survey was also sent to the IPPS SA members for participation. The face to face surveys were done on week days in the months of August and September 2019. A total of 55 completed surveys were obtained and the results analysed.

RESULTS

The results of the pilot survey indicated that most of the respondents interviewed was females (73%) under the age of 50 (67%). Most of them indicated that they are keen garden enthusiasts. When asked about gardening by only making usc of indigenous plant species, exotic species or a mix of both indigenous and exotic species in the garden, most agreed that a mixture of indigenous and exotic plants can be used successfully in a garden. Seventy-four present of the respondents was against a complete exotic garden whereas 50% agreed and strongly agreed that a garden should only contain indigenous species.

Fig. 1 indicates the respondent's responses concerning the various aspects of the different plants preferred. The leaves, flowers, fruit and growth form as well as the entire plant was evaluated. According to the results obtained the most prominent characteristics of these species selected was their flowers and secondly the fruit, with the flowers of *C. scabiosifolia* the top selective and the fruit of *L. edulis* as the top fruit/seed selective.

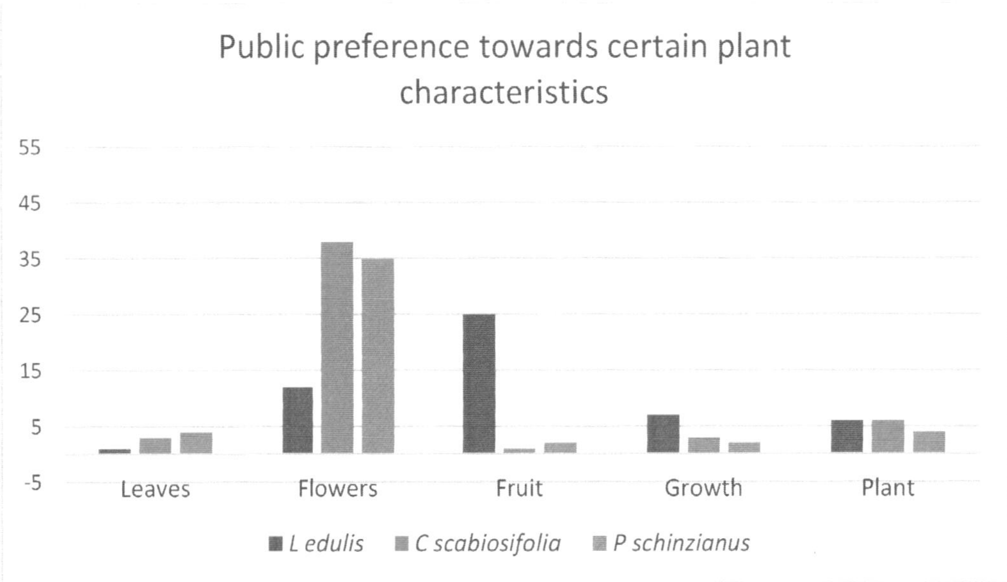

Figure. 1. A graphical representation of the respondent's responses towards popular features

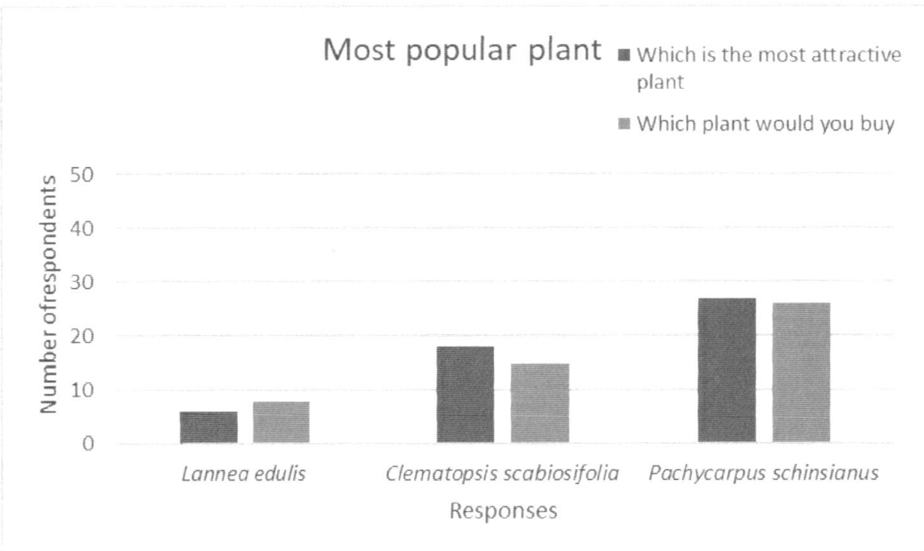

Figure 2. Graph indicating the plant species that the respondents found most attractive and the plant that they will most likely buy.

According to the results obtained the overall selection of the plant that the public would purchase and plant in their gardens was indicated as *P. schinsianus* (Fig. 2).

CONCLUSION

From the results obtained it is evident that it is necessary to proceed with a short pilot study to obtain the necessary feedback from the public concerning their preferences toward certain plant species and traits for cultivation before the actual cultivation and propagation research begin. These pilot studies should be extended to more garden centres and more response should be obtained to enable a more comprehensive assumption of what is being preferred by the public. These pilot studies can also create an awareness with the public towards indigenous species found in the Highveld Grassland. This literature review also emphasised the urgency of this project as only 0,5% of the Highveld Grassland is being conserved and many potential ornamental horticulture and eco-landscape species can be lost for future generations if proactive measurements are not being implemented.

Literature Cited

Bond, W.J. (2016). Ancient grasslands at risk: High biodiverse tropical grasslands are at risk for forest-planting efforts. Science *351*(6269). pp 120-122.

Bowie, R. and A. Frank. (2020). Highveld Grasslands. World Wildlife Fund. Washington, DC. http://www.worldwildlife.org/ecoregions/at1009 (access 12/03/2020).

Koekemoer, M, H.M. Steyn, and S.P. Bester. (2013). Guide to Plant Families of southern Africa. Strelitzia 31. South African National Biodiversity Institute, Pretoria.

Mucina, L. and M.C. Rutherford. (2006). The vegetation of South Africa, Lesotho and Swaziland. Strelitzia 19. SANBI. Pretoria.

Palmer, A.R. and A. Ainslie. (2005). Chapter 3: Grasslands of South Africa. In Suttie, J.M., Reynolds, S.G. & Batello, C. (Ed.) Grasslands of the world. Food and Agricultural Organisation of the United Nations. Plant Production and Protection Series no 34. [hppt://fao.org/3/y8344e00.htm](access 11/03/2020).

Van Wyk, B.-E. and N. Gericke. (2000). People's plants: a guide to useful plants of southern Africa. Briza Publications, Pretoria.

Wentzel, J. (2020). The "Wildflower Wholesale Nursery" Wild flower indigenous Plant Library. Lifestyle Pro Theme on Genesis Framework Word Press. [www.wildflowernursery.co.za] (access 12/03/2020).

Combined Proceedings IPPS 70:6-14. 2020.

The Future African Foraging Garden; an Experiment in African Urban Agriculture and Crop Conservation.

Jason Sampson

Department of Plant and Soil Sciences, Manie van der Schijff Botanical Gardens, University of Pretoria, Pretoria, 0002, South Africa

jason.sampson@up.ac.za
www.up.ac.za/botanical-garden
www.facebook.com/MvdSBG

Keywords: Food crops, indigenous, orphan crops, community outreach

Summary

Focused on African Orphan Crops, the Future Africa foraging landscape is part of the ongoing collaborative project that is the raison d'être for the establishment of this research orientated campus. This is a short synopsis of the talk given to the IPPS Southern Africa 2020 Annual conference on the history of the FA campus gardens. Subjects dealt with include the projects trajectory from inception to completion, the philosophy and practical aspects of its design and the future prospects of the concept and collection.

INTRODUCTION

The Future Africa project is situated on the old Experimental Farm of the University of Pretoria and forms part of the repositioning of the EF as the" Innovation Africa @ UP initiative".

Figure 1. An early morning photograph taken near the residences. Photo Credit Eyescape Photography.

Initially it was the brainchild of the Forestry and Biotechnology Institute (FABI), while the gardens are a collaborative effort with the Department of Plant and Soil Sciences (DPSS), partially due to their involvement in the Experimental Farm management and utilization but also due to other extremely successful hybrid sustainability/landscaping projects such as the rainwater harvesting garden on Hatfield Campus.

As the senior curator of the Manie van der Schijff Botanical Garden and the living plant collections of UP, I was responsible for managing the Departments interests in this regard and was involved in the project from inception, I was also responsible for its theme as well as composition.

Inception and ethos

Future Africa is envisioned as a hub for African and global research networks to address the challenges that hamper transformation towards a prosperous, equitable and sustainable future in Africa.

Its stated goals include:

- Fostering a new generation of science leaders
- Providing a dynamic living, learning and research environment
- Forming the nucleus of a community of scholars and other societal role players to advance excellence in scholarship, dialogue and impact to address specifically but not exclusively African problems

- Leverage the advantages of trans-disciplinary research to achieve these goals

The entire development consists of conference facilities, accommodation, residences, kitchens and the garden.

Design Philosophy

The Department of Plant and Soil Sciences and the Manie van der Schijff Botanical Garden became involved in the project in early 2015, close enough to the public announcement of the African Orphan Crops Consortium, a huge research network sponsored by the likes of the Mars Corporation and the WWF.

Orphan Crops can be defined as: socio-culturally relevant crop and trees grown in Africa which, historically or currently, may be side-lined by more "mainstream" crops although they may remain locally economically important. Orphan crops are hugely important in diversifying the food supply of humans and animals in Africa, and guarding against famine. These species may be indigenous to the continent or have become important to local peoples through trade and historical introduction.

Figure 2.
A panorama of the entire Future African Campus. Photo credit: Wynand Steyn.

The idea for the gardens theme was born out of this idea, and grew from the initial list of 101 "orphan African crops" issued by the AOCC:

It incorporates "orphan crops", listed or unlisted in the consortium list, as well as South African indigenous wild food plants. Humans evolved, initially in Africa, as hunter gatherers, and there is a huge array of uncultivated, wild food plants available in Southern Africa available to a collection such as this. Foraging eating, or harvesting wild plants for culinary use, or "eating the weeds and other things too" is becoming a huge cultural movement worldwide, even a conservation issue in certain cases and this development features an onsite kitchen with chef. Collaborations with, and research by the Consumer Sciences Department becomes a reality with such a collection on site.

Collections like this enable research into a host of different disciplines, from ennoblement of new food crops to genomic research but I have always envisioned it as focused on the ability of the urban landscape to provide food for fauna and people, and in so doing highlight not only the potentials that South African plants but also African plants can provide.

The Future Africa landscape will provide a living and functional testament that a predominantly indigenous landscape can be used as a food resource, this resource can also be utilised to teach, perform research and function as an aesthetic entity, all in one.

Species lists

A list of approximately 200 species of fruits, nuts, leafy greens, herbs, tubers etc., some wild foods, other crops was drawn up for inclusion in the project.

Some are exotics because they are AOC or other reasons, but the landscape is 90%+ indigenous / African, their inclusion based on traditional use as well as research potential but always considering their aesthetics. The landscape must be functional but pretty too!

INSPIRATIONS, LESSONS, AND INCLUSIONS

Rainwater Harvesting Garden on Hatfield Campus was a huge inspiration. The Attenuation dam on the FA site is also a rain water harvesting dam and runoff from roofs etc. end up in this water body, which is used for irrigation. The dam is used to grow water chestnuts (*Eleocharis dulcis*), waterblommetjies (*Aponogeton distachyos*), blue lotus (*Nymphaea caerulea*) as well as being stocked with Tilapia. The Green Walls of the Plant Sciences Building: On site we have used many "green facades", planted with species such as *Lablab purpurea* and *Mondia whitei*, for screening as well as shading.

Vertical gardening: still to be fully incorporated. We are soon to be hosting a PhD project on vertical OC farming.

Rehabilitation of the Hartebeespruit (ongoing): The Biodiversity Garden of Dr. Ida Breed is actively working towards making local biodiversity more acceptable to gardeners and landscapers. This is the only section on the ground which is not composed of edible planting.

The Cycad Collection of the MvdSBG is represented on site as a flagship planting of tropical African cycads around the baobab trees. These are not considered as edible plants due to a host of toxic principles, not to mention their value, but these plants were used as a food resource by indigenous hunter-gatherer peoples, after a host of preparation techniques, such as month's long fermentation. Definitely not recommended, but gorgeous plants in the landscape, nonetheless. Recycling and sustainability initiatives of the University of Pretoria such as the composting facility on our sports campus, both take the green waste of FA, but supply the composted by-product back.

Design Work

Building design and management of the project was performed by Earthworld Architects. Their vision can be described as "AFRI-TECH – Combining high level design processes with local resources and skills". "From the outset, the intention was to challenge existing design & construction processes. Each program would be addressed through a solution specific to each set of conditions."
(https://www.futureafrica.science/index.php/campus/future-africa-design/future-africa-buildings and https://www.ew-arch.co.za/post/3096/futureafrica/)

Insite Group Landscape Architects handled garden layout and specification: "Our heritage and indigenous natural resources can be used to better lives by capitalising on existing resources." (https://www.futureafrica.science/index.php/campus/future-africa-design/future-africa-gardens and http://insite-group.co.za/)
Design Team included Jason Sampson, Neal Dunstan (the then Landscape Architect of the University of Pretoria) and Philip Rousseau (the then junior Curator of the Cycad Collection, now sadly deceased).

Advantages of the site

These were multiple, the site is on the Northern aspect of a kopje, facing the Magaliesburg, microclimates between building, multiple grades allowing for specialised planting sites and good, fertile soil augmented by the composting facility of UP.
There is also a high maintenance budget which allows for a dedicated junior Curator, Ms. Lina Rampora, and the collection is considered a Heritage Collection of the University of Pretoria in its own right.

On site kitchens with dedicated Chef; this cannot be overemphasised. Meals and banquets are prepared out of the gardens regularly, and selected produce used almost every day.

Figure 3.
Looking West between residence blocks. This alleyway hosts the critically endangered Pondoland Coconut (*Jubeaopsis caffra*) specimens. Photo Credit: Eyescape Photography.

Figure 4. A southern view showing some of the large planting spaces for annual crops. Photo credit: Eyescape photography.

Selected species

Sterculia murex: The lowveld chestnut. Produces oil rich, sweet nuts in late summer. Edible root as well. Almost unknown outside of its restricted range around Nelspruit/Mbombela and surrounds.

Figure 5.
Sterculia murex, the Lowveld Chestnut. Entirely undomesticated, the large, oily seeds can be roasted in much the same way as true chestnuts (*Castanea*). Photos by author.

Tylosema esculent: Marama bean. Another completely wild food plant. Shows great ennoblement potential.

The broom cluster figs: *Ficus sur* and *Ficus sycamorus* the latter is actually the fig tree mentioned in the Christian Bible and was a sacred plant to the Ancient Egyptians. Both these plants produce multiple crops a year, the figs can be used green as pickles in the Indian fashion, or as preserves.

Mondia whitei: 'uMondi' a traditional African medicinal plant, it has edible leaves and bark which contains a compound related to vanillin. This is a rampant forest liana, and will cover a façade in a matter of months.

Figure 6. *Mondia whiteii* flowers, photo by author

Plectranthus esculentus: Also known as the Livingstone potato. Has been considered to be potentially locally extinct in South Africa. The large root clusters taste like minty potatoes and are extremely nutritious. Still cultivated as a subsistence crop in Zimbabwe.

Oxalis pes-caprae: A traditional food plant and food flavouring plant (herb) in the SW Cape.

Cucumis metuliferus: Farmed under the trade name of "Kiwano Melon" in New Zealand. Tastes better green (IMO), a sour cucumber substitute. Ripe melons are attractive but a bit bland.

Garcinia livingstonei: African Mangosteen. One of the best tasting wild fruit that the author has had the pleasure of sampling.

Portulacaria afra: The ever trendy 'Spekboom', the clone we have at Future Africa has a nice, soft textured leaf with a thin epidermis. Not at all bad in salad. The leaf is at its best taste in the morning due to malic acid concentration declining during the day.

m'Shaina*: Brassica* sp. or Venda mustard spinach, an heirloom green vegetable much consumed by the isiVenda people. A winter crop, this is probably the best tasting leafy green the author has ever eaten. I can recommend it highly.

Salvia 'Bee's Blue': a hybrid between *Salvia africana-lutea* and *S. dolomitica*, this indigenous sage has exactly the same flavour as S. officianalis, "proper" culinary sage.

Aponogeton distachyos: "Waterblommetjies, a South African veld food favourite. Still completely undomesticated, is grown in the attenuation dam.

FOR THE FUTURE

- Signage and interpretive routes
- Research
- Increase of collections
- Outreach and community engagement
- A cook book or three

Literature

National Research Council "Lost crops of Africa: volume I through 3", National Academies Press.

Palgrave, K. C. (2000). Trees of South Africa. Struik Publishers. Cape Town, Johannesburg.

Van Wyk, B., and N. Gericke. (2000). People's plants: A guide to useful plants of Southern Africa. Briza publications.

Van Wyk, B. (2005). Food plants of the world. Timber Press.

Combined Proceedings IPPS 70:15-21. 2020.

Year 2020 Summary -The International Plant Propagators' Society-Southern Region of North America - (SRNA) Virtual Meetings

Brie Arthur

Brie Grows, B.A. Communications
7624 Troy Stone Drive, Fuquay-Varina, NC 27526, USA

brienne.gluvna@gmail.com

Keywords: Annual Meeting, CoVid-19, North American Virtual Summit (NAVS), Southern Region of North America (SRNA), Parkerson Virtual Student Research Competition

Summary

Due to the pandemic, there was no annual meeting in 2020 of the IPPS- Southern Region of North America (SRNA). In place of the annual meeting, the SRNA successfully led, planned and executed the 3-day North American Virtual Summit (NAVS) during Oct 27-29, 2020. There was a total of a total of 947 attendees. There were three virtual presentations given by Drs. David Creech, Dennis Werner and Mike Dirr - as well as virtual nursery tours. Nine students competed in the Charlie Parkerson Virtual Student Research Competition. Three students delivered 20-min presentations and the remaining six students gave shorter, "oral poster "presentations. The SRNA remains in good financial condition thanks to the conscientious management of Sec-Tres, Donna Foster. Dr. Gary Knox received the Meadows Award and Dr. Dave Creech was recognized as a Fellow. The SRNA currently plans to hold the 45th Annual meeting on Oct 23-26, 2021 in Mobile, Alabama.

PRESIDENT BRIE ARTHUR

Because of the Covid-19 Pandemic there was no 45th Annual Meeting of the IPPS-SRNA). At the June 2020 mid-year EC Board meeting - the difficult decision was made to cancel the 2020 annual meeting, planned for Tulsa. It also made sense to retain the same EC board and all committee assignments through 2021, since there was no annual meeting (Fig. 1).

Figure 1. President Brie Arthur (left) with Past-President Elliott Hallum (right).

NORTH AMERICAN VIRTUAL SUMMIT (NAVS)

In place of the annual meeting, the SRNA successfully led, planned, and executed the 3-day North American Virtual Summit (NAVS) during Oct 27-29, 2020. Key players in organizing the Southern Region NAVS were SRNA President - Brie Arthur, Dr. Cheryl Boyer (hosted the Zoom conferencing from Kansas State University), Laura Barth (SR digital communications manager), Donna Foster (Sec-Tres) and Bobby Green, first Vice President. The NAVS committee included members from the Eastern and Western Regions, which also had their 2020 annual conferences cancelled. The format was 3 half-day virtual sessions, with each region scheduling their own programs, including educational speakers and nursery tours. There was great effort that went into the logistics and execution of the NAVS – which was a new first for the IPPS.

The NAVS was a great success. There were 902 people registered (it was at no cost this year), plus 45 panelists for a total of 947 attendees. Equally important, there were 343 non-IPPS members who attended (a chance to recruit new members) from 15 different countries.

For the SRNA virtual session, presentations given by Drs. David Creech, Dennis Werner and Mike Dirr.

TIPS AND TRICKS FOR PROPAGATING DIFFICULT TO ROOT SPECIES - David Creech, Director SFA Gardens, Stephen F Austin State University, Nacogdoches, TX (Fig. 2).

A lifetime of sticking a wide range of woody and herbaceous cuttings was condensed into twenty minutes. Every batch of cuttings is a brand-new story and a myriad of variables influence successful rooting. Time of year, hormone application (powder, solution, or gel), mist propagation, bottom heat, heated mist, substrate drainage, light (intensity and photoperiod) - and the character of each cutting is a determinant. For difficult to root species, the best propagators are constantly experimenting to find the exactly right timing and protocol that leads to successful rooting.

http://www.ipps.org/proceedings-volume/North-American-Summit-2020/194#/video/6

Figure 2. Dr. David Creech, Director Stephen F. Austin Gardens, Regent's Professor Emeritus, Stephen F Austin State University, Nacogdoches, TX.

DESIGNER GENES, DESIGNER REDBUDS: TWENTY YEARS OF *CERCIS* BREEDING - Dr. Dennis Werner, J.C. Raulston Distinguished Professor Emeritus, JC Raulston Arboretum, North Carolina State University (Fig. 3.).

Werner summarized the accomplishments made at the North Carolina State University redbud breeding program during the past 20 years. He reviewed the most recent cultivar releases, and discussed the goals of the program moving forward. He also discussed some of the fascinating mutations in redbud, and how they interact biologically to allow for the creation of new and novel forms. Dr. Werner's redbud selections include: 'Ruby Falls', 'Merlot', 'Whitewater', 'Pink Pom Poms', Flame Thrower, and Golden Falls. His numerous cultivar releases in Buddleia have focused on sterility, compact growth habit, and expansion of the color palette. Dr. Werner's presentation was only available to the live audience.

Figure. 3. Dr. Dennis Werner, JC Raulston Arboretum, North Carolina State University, Raleigh, NC

TOUR OF THE DIRR GARDEN AND PREMIER INTRODUCTIONS, INC.: BEST PERFORMERS AND THOSE TO COME - Dr. Michael A. Dirr, President, Chestnut Oak Farm and partner in Premier Introductions, Inc., Athens, GA, (Fig. 4).

Legendary plantsman Michael Dirr shared his insights on new plants from his home garden and breeding facility, Premier Introductions, Inc. Dirr's Georgia Plant Introduction Program has introduced over 40 new cultivars into the nursery trades. He remains active in research and new plant development.

http://www.ipps.org/proceedings-volume/North-American-Summit-2020/194#/video/7

Figure 4. Dr. Michael A. Dirr, President, Chestnut Oak Farm and partner in Premier Introductions, Inc. Athens, GA.

CHARLIE PARKERSON VIRTUAL STUDENT RESEARCH COMPETITION

Prior to the NAVS, the Virtual Charlie Parkerson Student Research Competition was held via zoom (Fig. 5). There were nine students who participated in the Charlie Parkerson Virtual Student Research Competition The competition was led by Chair, Mack Thetford, Cheryl Boyer, Jim Robbins and Brie Arthur. http://www.ipps.org/proceedings-volume/North-American-Summit-2020/194#/video/10

There were 65 participants on the zoom. The top three student finalists in the competition gave 20-min oral presentations. The remaining six students gave shorter, 'oral poster" presentations via zoom. There were eight papers published from the student competition in the 2020 *IPPS Combined Proceedings.*

The link for the presentations of the Charlie Parkerson Virtual Student Research Competition is

http://www.ipps.org/proceedings-volume/North-American-Summit-2020/194#/video/11

The three finalists in the order of their placement in the competition were:

1) S. Brooks Parrish, Renjuan Qian[2], Sandra B. Wilson[3], and Zhanao Deng. 2020. **Morphological and Cytological Characterization of Six Porterweed (*Stachytarpheta* spp.) Selections.** University of Florida.
2) Yuvraj Khamare, S. Christopher Marble[2], James E Altland[3] and Annette Chandler. 2020. **Influence of Substrate Stratification and Fertilizer Placement on Growth of Ligustrum (*Ligustrum japonicum*) and Germination and Biomass of Bittercress (*Cardamine flexuosa*) in Containers.** University of Florida.
3) Brian A. Schulker, Brian E. Jackson, and William C. Fonteno. 2010. **Where Did theWater Go?** Department of Horticulture, North Carolina State University.

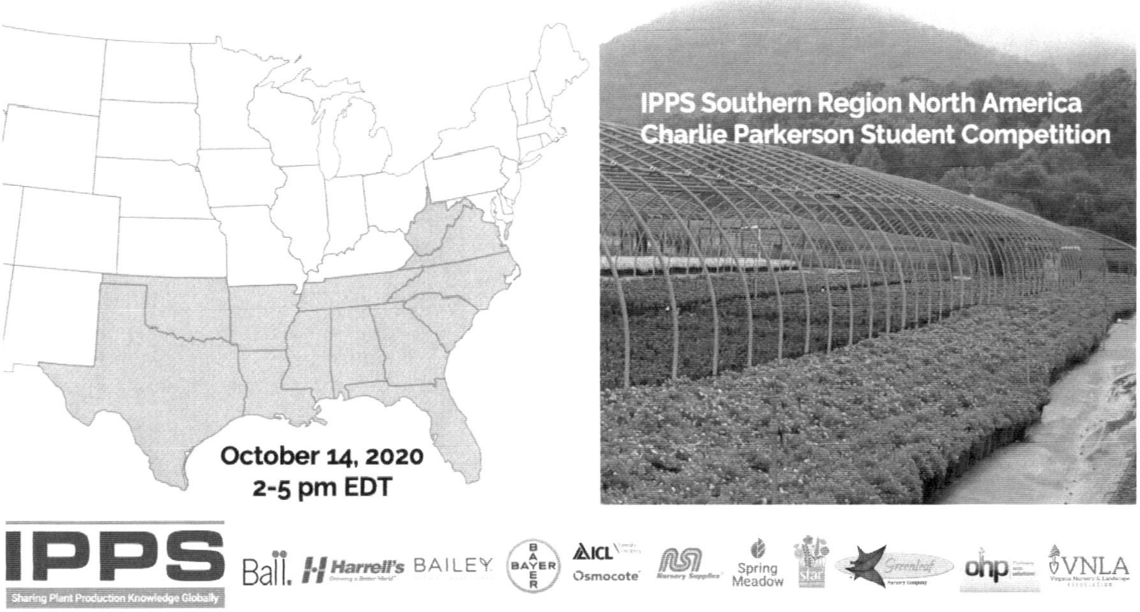

Figure 5. The Charlie Parkerson Student Competition and IPPS-SRNA sponsors.

OTHER HIGHLIGHTS OF 2020 IPPS-SRNA

- There was also a virtual Question Box, moderated by Dr. Judson LeCompte. International Director, Tom Saunders, complemented LeCompte for the excellent, "very smooth" job he did in moderating the virtual Question Box. From Judson's cool digs in Michigan, it was like an FDR fireside chat!
- Dr. Gary Knox of the University of Florida was recognized with the Meadows Award for 2020 (Fig 6.).

Figure 6. Dr. Gary Knox of the University of Florida, Meadows Awardee for 2020.

- Dr. David Creech of Stephen F. Austin University was recognized with the Fellows Award for 2020 (Fig. 7).

- To honor Margie Jenkins, the "Margie Jenkins Student Conference Scholarship" is being established. There is a committee currently working on the details. This is in addition to the existing "Vivian Munday Student Mentor Conference Scholarship".

Figure 7. Dr. David Creech, Fellows Awardee for 2020.

SRNA-IPPS BUSINESS UPDATE

- The brand-refresh (name change) to "IPPS: International Plant Production Society" (rather than Propagators) was narrowly defeated: *63.6% in favor of the name change with a 2/3rds (67%) approval needed.* The SR considers the brand-refresh critical for being more inclusive in attracting new members, marketing, and garnering sponsorship to run the organization. At the Oct 2019 SR annual business meeting, there was a unanimous vote for the name change (60-0). Of the 77 SR members who recently voted in the International Web solicitation, 10 voted against it; this was the difference in the brand-refresh not being approved. Of the 8 regions who voted: five of the eight voted by 2/3rds or more - for the name-change. New Zealand is the only region (48% members approved) – that does not have a majority in favor. It is strategically important that the IPPS move forward with the brand-refresh.
- The SR Education Endowment Fund administered by HRI had a balance of $71,832 as of 2 November 2020. We are ahead of the initial $60,000 goal, but want to continue building the Endowment more to support SRNA Educational programs: *Early-Career Propagator*

Exchange program, Parkerson Student Research competition, Vivian Munday Young Horticultural Professional Scholarship Work Program, and help subsidize annual registration costs. Keven Gantt chairs the Endowment Fund, which was started four years ago with an anonymous $20,000 contribution.

- The 2020 SRNA has 239 paid members and 28 student members for a total of 267 members; considering we did not have an annual meeting, the SRNA is holding its own with membership retention & growth. 2019: 245 paid + 18 student members; 2018: 258 paid members + 13 student members; 2017: 253 paid members; 2014: 225 paid members.
- President Brie Arthur has been chairing the combined monthly meetings of the membership committee and technical communications committee. They review content to include in the newsletter, creative strategies for gaining new membership, video content, etc. In the future the SRNA may hire a videographer to do 5-10 min segments of nurseries, garden centers, gardens for the SR-You-Tube channel. The EC is keen on using You-Tube videos to market the SRNA.
- The SRNA is in good financial shape, thanks to the conscientious management of Sec-Tres, Donna Foster, guidance of previous and current EC boards, and major efforts to garner industry sponsorship. Cancelling the 2020 planned annual meeting saved considerable hotel and meeting losses that would have occurred. As of 31 October, the SRNA balance was. $195,882.
- As of 24 November 2020, the SRNA is still planning for the Oct 2021 Mobile Meeting to take place. For the 2nd year in a row, Program Chair Bobby Green is preparing with a slate of excellent, entertaining speakers for the Mobile conference!

Combined Proceedings IPPS 70:22-28. 2020.

Morphological and Cytological Characterization of Six Porter-weed (*Stachytarpheta* spp.) Selections

S. Brooks Parrish[1a], Renjuan Qian[2], Sandra B. Wilson[3], and Zhanao Deng[1]

[1]University of Florida, Department of Environmental Horticulture, Gulf Coast Research and Education Center, 14625 County Road 672, Wimauma, FL 33598; [2]Zhejiang Institute of Subtropical Crops, 334 Xueshan Road, Wenzhou, Zhejiang 325005, China; [3]University of Florida, Department of Environmental Horticulture, P.O. Box 110675, Gainesville, FL 32611

brooks.parrish@ufl.edu

[a]First Place- Charlie Parkerson Graduate Student Research Paper Competition

Keywords: Chromosome, DNA content, growth habit, ploidy.

Summary

Porterweed (*Stachytarpheta* spp.), a member of the verbena family, is a common ornamental plant in warmer parts of the U.S. that is frequently used in pollinator gardens to attract many species of butterflies and hummingbirds. Much floral diversity exists within the genus and hybrid forms. This study was conducted to assess the growth habit, flowering, DNA content, and chromosome number of six porterweed selections to explore the relationship among species. Results identified three distinct porterweed growth habits (upright, semi-upright, and prostrate) and showed that nuclear DNA content ranged from 2.95 to 3.79 pg/2C. Chromosome counting revealed that all porterweed accessions tested were polyploid (tetraploid, pentaploid and hexaploid), with the exception of dwarf blue porterweed (*Stachytarpheta* spp) that was darkly stained chromosomes as they become organized in the metaphase stage of cell division. Subsequent cytological and morphological comparisons can be used to not only readily distinguish invasive and non-invasive forms of porterweed, but aid in future breeding programs.

IPPS Vol. 70 – 2020

INTRODUCTION

Porterweed (*Stachytarpheta* spp.) is an ornamental that is desired for its abundant brightly colored flowers that attract a diverse array of pollinators throughout much of the year. It is a drought tolerant, low maintenance plant commonly used in the southern United States as a perennial in warmer climates (USDA cold hardiness 9a), an annual in more temperate climates, or in container gardens. The *Stachytarpheta* genus is vast with 133 species identified in Australia (Munir, 1992) and 79 species classified in Brazil (Atkins, 2005). Seed is readily available to the public and can be found from many nurseries and online sellers.

In Florida, the most commonly sold porterweed species are jamaican porterweed (*Stachytarpheta jamaicensis),* nettleleaf porterweed (*Stachytarpheta cayennensis),* coral porterweed *(Stachytarpheta mutabilis),* purple porterweed (*Stachytarpheta frantzii*) and dwarf red porterweed (*Stachytarpheta sanguinea*). Jamaican porterweed is native to dunes, shell middens, pine rocklands, and disturbed sites of the central and southern Florida peninsula (Wunderlin and Hansen, 2020); whereas, nettleleaf porterweed was introduced to the United States from Central and South America and has escaped cultivation. While it has not yet altered native plant communities in Florida, nettleleaf porterweed is listed as a Category II invasive plant by the Florida Exotic Plant Pest Council (FLEPPC) due to its increased abundance or frequency (FLEPPC, 2019). The University of Florida Institute of Food and Agricultural Sciences Status Assessment of Non-native Plants recommends its "use with caution" (UF/IFAS Assessment, 2020). Hybridization potential between the native and invasive congeners is of concern.

In prior work, Wilson et al. (2009) evaluated seed production and viability of eight porterweed selections in Florida and found 'Violacea' (*S. mutabilis*), 'Naples Lilac' (*S. cayennensis* × *S. mutabilis* 'Violacea') and 'Mario Pollsa' (*Stachytarpheta* spp.) porterweed to be highly female sterile. Also in their study, through controlled manual crosses, the potential for *S. cayennensis* to hybridize with × *S. jamaicensis* was realized.

Chromosome number and ploidy level are important plant characteristics, and the latter is an important factor determining hybridization potential. Fedorov (1974) and Sanders (2001) reported that the porterweed genus has varying numbers of chromosomes from $2n$ = 18 to $2n$ =160 and varying levels of ploidy. However, these reports did not publish chromosome images likely due to low quality resolution. The use of acids for cell wall degradation and stains such as crystal violet or acetocarmine for staining have been popular choices for chromosome counting, but sometimes lack clarity with certain samples (Dalgaard, 1986). When plant cells contain large numbers of chromosomes, it is especially important and critical to have effective chromosome squashing techniques that can make chromosomes well spread and produce clear images of chromosomes.

The purpose of this study was to characterize growth habits and cytological features of six porterweed selections. The main objective was to confirm the ploidy level of common porterweed cultivars by chromosome counting and understanding the relationship between chromosome number and nuclear DNA content determined by flow cytometry. This information is critical for future porterweed plant breeding programs, and also for the identification of hybrids. Ornamental plant breeders could benefit greatly from a reliable chromosome squashing protocol that will produce high quality metaphase images.

MATERIALS AND METHODS

Plant materials. Six porterweed selections were evaluated in this study. Jamaican, coral, nettleleaf, and 'Naples Lilac' porterweed plants were obtained from a previous study conducted by Wilson et al. (2009). U*J 3-2 resulted from manual crossing of *S. cayennensis* and *S. jamaicensis* in a greenhouse located at the Gulf Coast Research and Education Center (Wimauma, FL). Dwarf blue porterweed plants were obtained from Grandiflora Nursery, Inc. (Gainesville, FL). Vegetatively propagated porterweed plants were grown in gallon plastic containers filled with Fafard 2P mix (Florida Potting Soil, Orlando, FL).

Growth Habit and Flowering. Fully mature flowering plants were used to assign categories of growth habit and flowering. Growth habit was identified as upright, semi-upright, or prostrate. Flower production was quantified as low and high, where high flowering plants had more than 60 florets on a single spike. Five replicates were accessed for each porterweed selection.

Determining nuclear DNA content. An Accuri C6 flow cytometer (BD Biosciences, San Jose, CA) was used to determine nuclear DNA content. The flow cytometry protocol recommended by Doležel et al. (2007) was followed using rye [*Secale cereal* 'Daňkovské' (16.19 pg·$2C^{-1}$)] as the internal standard. Three flow cytometrical analyses were run for each porterweed selection, and a minimum of 3000 nuclei were counted per run. Nuclear DNA content (pg/2C) was calculated according to Doležel et al. (2007).

Squashing and counting chromosomes. The chromosome squash protocol was adapted from Chen et al. (1982). Before 10:00 AM, vigorously growing root tips (1cm) were excised from porterweed plants and treated in 0.002 M 8-hydroxyquinoline for 4 h in the dark. Root tips were immersed in a fixative solution (3 methanol: 1 acetic acid, v/v) for at least 2 h. The fixed roots were rinsed three times in deionized water before a much smaller section of the root tips (approximately 1 mm) was excised and macerated in an enzyme solution containing 2.5% cellulase and 2.5% pectinase for 3 h 15 min inside an incubator at 27 °C. Macerated root tips were washed in deionized water for 10 min and then fixed in a fixative for 0.5 h. Root tips were squashed in a drop of the fixative solution on a pre-chilled microscopic glass slide. The prepared slide was heated over an alcohol burner for a few seconds and stained with a 2.5% Giemsa solution (Sigma, St. Louis, MO) for 10 min. Stained glass slides were rinsed in distilled water, air-dried, and then observed at 1000× magnification under a BX41 microscope with an Olympus Q-color 5 camera (Olympus America Inc., Melville, NY).

RESULTS

Three growth habit categories were identified from the six porterweed selections. Coral and 'Naples Lilac' grew upright; nettleleaf and dwarf blue grew semi-upright, and jamaican and U*J3-2 porterweed grew prostrate (Fig. 1). Nettleleaf, U*J3-2, and jamaican porterweed recorded 65-70 flowers at the time of evaluation. Coral, 'Naples Lilac', and dwarf blue porterweed had much fewer flowers with only 10-30 flowers each. Based on plant growth rather than flower number, these results emphasize how the native jamaican porterweed can be phenotypically distinguished from the invasive nettleleaf porterweed.

Figure 1. Images of each porterweed accession at the time of data collection. Plants were propagated at the same time and grown under the same conditions.

Nuclear DNA content of the selections had a range of 0.98 pg. At least 31 metaphases were observed for each accession which revealed four chromosome numbers (*2n* = 42, 112, 140, and 168) (Table 1). Chromosome analysis categorized accessions from diploids to hexaploids with two different base chromosome numbers (Table 1).

Table 1. Nuclear DNA content and chromosome number of six porterweed selections.

Taxa	Species	Nuclear DNA content ± SD (pg/2C)	Metaphases observed	Chromosome number (2n)	Base chromosome number	Ploidy level
Coral	*S. mutabilis*	3.66 ± 0.05	38	168	28	6x
Naples Lilac	*S. cayennensis* × *S. mutabilis* 'Violacea'	3.79 ± 0.04	41	168	28	6x
Nettleleaf	*S. cayennensis*	2.81 ± 0.03	36	112	28	4x
U*J 3-2	*S. cayennensis* × *S. jamaicensis*	3.28 ± 0.05	36	140	28	5x
Jamaican	*S. jamaicensis*	3.73 ± 0.09	31	168	28	6x
Dwarf Blue	*S.* spp.	2.95 ± 0.03	35	42	21	2x

The ploidy level of the native Jamaican porterweed was 6x, whereas the ploidy level of the invasive nettleleaf porterweed was 4x. Images of chromosomes of each accession can be clearly seen in Figure 2. Results from this study not only serve as a first report of chromosome numbers of specific porterweed cultivars; but present a successful new technique to study chromosomes in other ornamental species.

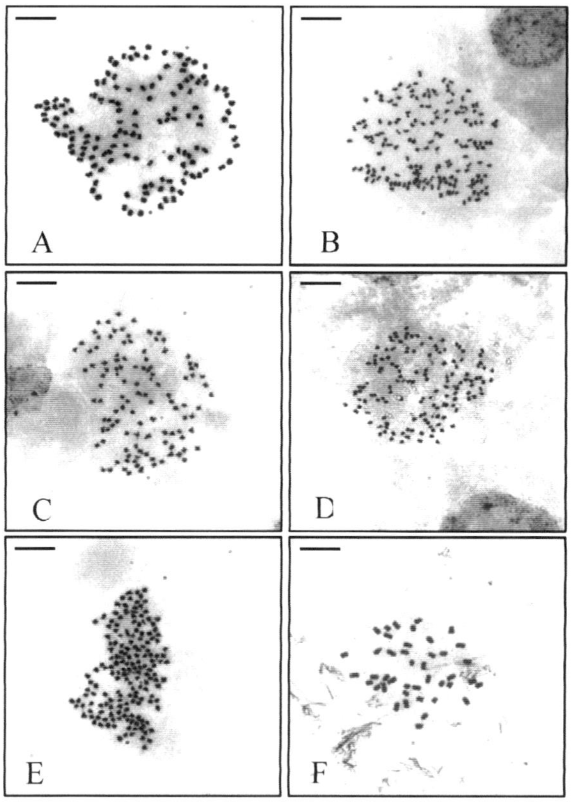

Figure 2. Micrographs (×1000) of somatic chromosomes observed in root tip cells stained in giemsa. A: coral porterweed ($2n$ = 168), B: 'Naples Lilac' porterweed ($2n$ = 168), C: nettleleaf porterweed ($2n$ = 112), D: U*J3-2 porterweed($2n$ = 140), E: jamaican porterweed ($2n$ = 168), and F: dwarf blue porterweed($2n$ = 42); scale bar = 10 µm.

DISCUSSION

All porterweed accessions evaluated in this study had purple to violet flower colors, with the exception of coral porterweed which had dark pinkish colored flowers (Fig. 3). With only small differences in flower color, the growth habit of each porterweed selection proved to be a feature that could aid in their identification. It is of interest to note that U*J3-2 took on the prostrate growth habit of its male parent (jamaican porterweed) but was considerably wider than both parents. A similar segregation of traits was observed in a hybridization of *S. angustifolia* × *S. cayennensis* produced by Solanke et al. (2019).

Flow cytometry yielded sharp peaks for both the internal standard and each of our samples. Standard deviation (SD) values for the mean nuclear DNA content were ≤ 0.09 pg (Table 1). Nuclear DNA content for the accessions ranged from 2.81 pg/2C in nettleleaf porterweed to 3.79 pg/2C in 'Naples Lilac'. This is of interest as 'Naples Lilac' is reported to be a cross between the invasive nettleleaf porterweed and the non-invasive 'Violacea' porterweed (*Stachytarpheta mutabilis*) (Kastenholz, personal communication).

A total of 217 cells in the metaphase stage of cell division were observed, photographed, and counted to determine the somatic chromosome number for all six selections. The stained cells produced had exceptional clarity that allowed the precise counting of the many small chromosomes. Surprisingly, dwarf blue porterweed recorded a higher average nuclear DNA content than nettleleaf porterweed, but had less than half the number of chromosomes, differing by 70 chromosomes. The identification of polyploids will be essential for breeders to use in generating non-invasive sterile cultivars. Sterile cultivars of *Lantana camara* have successfully been developed by the crossing of tetraploid and triploid cultivars (Czarnecki et al., 2014). Wilson et al. (2009) identified 'J.P's Pink' (*S. speciosa*) and 'Red Compact' (*S. speciosa*) porterweed as diploid species, making them potential candidates for crosses with now identified tetraploid nettleleaf porterweed. Furthermore, confirming polyploids through chromosome staining has laid the groundwork for determining the ploidy of other porterweed species with flow cytometry.

DNA content and chromosome counts presented for the six porterweed selections will undoubtably pave the way for the production of sterile porterweed cultivars as safe alternatives to the invasive form. The chromosome squashing protocol identified within has high potential for adaptation to other plant species. The technique requires few resources and can be completed in just two days yielding chromosome spreads with high clarity. Stained cells produced could further be used in chromosome measurements and forming karyotypes for plant species. Ornamental plant breeding programs often lack access to expensive genomic analysis. The use of this high throughput chromosome visualization protocol could be a major leap in the development of new ornamental hybrids.

Literature Cited

Atkins, S. (2005). The genus *Stachytarpheta* (Verbenaceae) in Brazil. Kew Bull. 60:161–272.

Czarnecki, D.M., Hershberger, A.J., Robacker C.D., Clark, D.G., and Deng, Z. (2014). Ploidy levels and pollen stainability of *Lantana camara* cultivars and breeding lines. HortScience. *49*:1271-1276.

Chen, R.Y., Song, W.Q., and Li, X.L. (1982). Wall degradation hypotonic method of preparing chromosome samples in plant and its significance in the cytogenetics. Acta Genetica Sinica. *9*:151–159.

Dalgaard, V. (1986). Chromosome studies in flowering plants from Macaronesia. Anales Jardín Botánico de Madrid. *43*:83-111.

Doležel, J., Greilhuber, J. and Suda. J. (2007). Estimation of nuclear DNA content in plants using flow cytometry. Nat. Protoc. *2*:2233–2244.

Fedorov, A. (1974). Chromosome numbers of flowering plants. Reprint ed. Otto Koeltz Science, Koenigstein, West Germany.

Florida Exotic Pest Plant Council. (2019). Florida exotic pest plant council's 2019 list of invasive plant species. 25 March 2020. https://www.fleppc.org .

Munir, A.A. (1992). A taxonomic revision of the genus *Stachytarpheta* Vahl (Verbenaceae) in Australia. J. Adelaide Botanic Garden. *14*:133–168.

Sanders, R.W. (2001). The genera of Verbenaceae in the southeastern United States. Harv. Pap. Bot. *5*:303–358.

Solanke, D.S., Oziegbe, M., and Azeez, S.O. (2019). Interspecific hybridization studies of three *Stachytarpheta* species from Nigeria. Jordan J. Biol. Sci. *12*:435-440.

University of Florida, Institute of Food and Agricultural Sciences (UF/IFAS). Assessment of Non-Native Plants in Florida's Natural Areas. (2020). *Stachytarpheta cayennensis*. 10 April 2020.
https://assessment.ifas.ufl.edu/assessments/stachytarpheta-cayennensis/ .

Wilson, S.D., Knox, G.W., Muller, K.L., Freyre, R., and Deng, Z. (2009). Seed production and viability of eight porterweed selections grown in northern and southern Florida. HortScience *44*:1842-1849.

Wunderlin, R. P., Hansen, B. F., Franck, A. R., and Essig, F. B. (2020). Atlas of Florida Plants (http://florida.plantatlas.usf.edu/). [S. M. Landry and K. N. Campbell (application development), USF Water Institute.] Institute for Systematic Botany, University of South Florida, Tampa.

Combined Proceedings IPPS 70:29-37. 2020.

Influence of Substrate Stratification and Fertilizer Placement on Growth of Ligustrum (*Ligustrum japonicum*) and Germination and Biomass of Bittercress (*Cardamine flexuosa*) in Containers

Yuvraj Khamare[1a], S. Christopher Marble[2], James E Altland[3] and Annette Chandler[4]
[1]Graduate Research Assistant; [2]Assistant Professor; [3]Research leader, [4]Biological Scientist III;[1,2,4]University of Florida/IFAS, Mid-Florida Research and Education Center, Apopka, FL, USA, [3]Application Technology Research Unit, USDA-ARS, Wooster, OH, USA.

ykhamare@ufl.edu

[a]Second Place- Charlie Parkerson Graduate Student Research Paper Competition

Keywords: Resource efficiency, weed control, weed management

Summary

Substrate stratification is a method of filling nursery containers with pine bark (or other substrates) with different particle sizes in "layers" in order to improve soil moisture dynamics. Currently, substrate stratification, or layering, is being investigated by some researchers as a method to increase the efficiency of production inputs such as irrigation and fertilization. Itis typically performed using larger particle bark as the bottom substrate and finer particle bark as the top substrate to achieve more uniform moisture distribution within containers. The objective of this study was to evaluate the effect of stratified substrates and strategic fertilizer placement on the growth of common nursery weeds and ornamental crops. In contrast to typical methodology, this study evaluated use of coarse bark (screened to 1.3 or 1.9 cm) as the top substrate and finer bark (0.95 cm) as the bottom substrate with the goal of reducing water holding capacity in the top 5 to 7.5 cm of the substrate to reduce weed germination and growth. Results showed that substrate stratification significantly decreased the growth of bittercress (*Cardamine flexuosa*) by 85% to 90% in comparison with substrates that were not stratified. While stratification initially reduced growth of ligustrum (*Ligustrum japonicum*), at 6-months after potting , there was no difference in ligustrum shoot or root weight in comparison with non-stratified industry standard substrates. The results indicate that substrate composition along with strategic fertilizer placement can be utilized as an effective weed management strategy.

INTRODUCTION

Nursery growers rely on frequent application of preemergence (PRE) herbicides along with supplemental hand weeding to manage weeds as there are no postemergence (POST) herbicides labeled to be safely applied on top of the ornamentals for broadleaf weed control. While PRE herbicides are effective, they are not labeled for use on all ornamental species as tropical plants, succulents, and herbaceous annuals and perennials are highly sensitive to herbicides. There are several negative consequences associated with frequent use of preemergence herbicides including high chemical costs, runoff and potential environmental effects, crop safety concerns, and inefficiency. For example, when applying herbicides to spaced containers, as much as 80% of the herbicide lands in between containers and is unavailable for weed control, resulting in sunk costs (Gilliam et al., 1992). Challenges associated with herbicide use often lead to the need for frequent hand weeding. However, hand weeding is often the costliest weed management strategy, costing up to $9,000 ha^{-1} (Case et al., 2005; Stewart et al., 2017). Due to the cost of hand weeding and the challenges associated with PRE herbicides, more integrated weed management strategies are needed for nursery production.

Strategic fertilizer placement has been investigated as an integrated weed management method in recent years. Alternative, or "strategic" fertilizer placements, such as subdressing so that the crop has access to nutrients but weeds do not, has been shown to reduce growth of common nursery weeds such as spotted spurge (*Euphorbia maculata*) and eclipta (*Eclipta prostrata*) by over 80% (Saha et al., 2019; Stewart et al., 2019). An additional area that has recently been investigated as a method of improving moisture distribution in nursery containers is substrate stratification, or layering, in which different portions of a container are filled with substrates of varying particle sizes (Fields et al., 2020). As gravitation potential decreases from the top to the bottom of a container, there is a gradient of increasing substrate moisture from the top to the bottom of the container (Owen and Altland, 2008). With stratified substrates, substrates could be layered within a container so that larger particle materials with lower water holding capacity were at the bottom of the container and finer textured materials with higher water holding capacity were at the top. Consequently, a more constant moisture gradient would be created with the benefit of conserving both water and nutrients (Fields et al., 2020).

Depending upon the composition and layering, stratified substrates could also potentially reduce weed growth if substrates were stratified in an inverse manner to that described by Fields et al. (2020), using larger particle materials in the top portion of the container profile. This composition would lead to a substrate that dried quickly on the top surface, reducing weed growth, but held adequate moisture for crop growth. An essential prerequisite of seed germination is water uptake by seed (Harper and Benton, 1966). Previous research has shown that germination and growth of many common nursery weeds such as pearlwort (*Sagina procumbens*), northern willowherb (*Epilobium ciliatum*), and common groundsel (*Senecio vulgaris*) decrease as substrate particle size increased (Wada, 2005). Additionally, a container substrate surface with a larger particle size could cause weed seeds to be flushed deep into the substrate, reducing, or eliminating germination because many weeds require light in order to germinate (Keddy and Constabel, 1986). However, the effects of substrate stratification, used either alone or combined with alternative fertilizer placement on growth of nursery weeds and ornamental crops is unknown. The objective of these experiments was to evaluate the effect of stratified substrates and strategic fertilizer placement on growth of common nursery weeds and ornamental crops.

MATERIALS AND METHODS

All experiments were conducted at the Mid-Florida Research and Education Center in Apopka, FL in November 2019 and repeated in February 2020. Pine bark was purchased from a local supplier and then screened through standard soil sieves to yield three different particle sizes including 0.95, 1.3, and 1.9 cm pine bark. Each resulting particle size included all bark particles that were equal to or smaller than each sieve size. The first three substrate treatments were constructed using one of the three particle sizes (0.95, 1.3, or 1.9 cm bark) throughout the container and had controlled release fertilizer (CRF) (Osmocote® 17-5-11, 8 to 9 mo.) incorporated at 35 g pot^{-1} throughout the substrate profile (TO) (0.95:TO, 1.3:TO, and 1.9:TO, respectively) following standard industry practices. Stratified substrate treatments were constructed by having either the 1.3 or 1.9 cm pine bark as the top substrate with the bottom substrate comprised of 0.95 cm bark. The top substrate (1.3 or 1.9 cm pine bark) was applied at depths of either 5 or 7.5 cm, resulting in four stratified (S) substrate treatments (1.3:S:5, 1.3:S:7.5, 1.9:S:5, and 1.9:S:7.5). In all stratified treatments, the CRF and rate mentioned previously was incorporated at the same rate per pot in all cases, but was only incorporated into the bottom 0.95 cm substrate. An additional treatment consisted of removing all fines from the 1.9 cm bark and using it as the top substrate applied at a 5-cm depth (1.9:S:5: N/F) as described previously.

This resulted in a total of eight substrate treatments with the 0.95:TO, 1.3:TO, and 1.9:TO considered industry standard substrates in that they were comprised of particle sizes often selected for use by growers and contained CRF incorporated throughout the substrate profile.

Uniform liners of ligustrum (*Ligustrum japonicum*) grown in 5-cm plug trays were used to assess the response of a common ornamental to the stratified substrate treatments. During transplanting, liners were planted using standard planting methods in 3.8 L nursery containers, and the top substrates were not applied as a mulch would be applied. That is, the roots of the ligustrum were planted into the top portion of all substrates. After ligustrum were transplanted into the above-mentioned substrates, all plants were placed on a full sun nursery pad, irrigated 1.3 cm per day via overhead irrigation, and were evaluated for 6 months after transplanting (MAT). Data collected included plant growth index [(height + width at widest point + perpendicular width) ÷ 3] measured every 2 months, in addition to root and shoot dry weights at study conclusion. The first experimental run was initiated on Nov. 20, 2019 and the second on Feb. 2, 2020.

To assess weed growth, 25 seeds of bittercress (*Cardamine flexuosa*) were surface sown into a separate set of 3.8 L nursery containers that were filled and fertilized as mentioned previously, placed inside a shade house (60% ambient light), and were irrigated 1-cm per day via overhead irrigation. Data collection included counts of emerged bittercress after 4 and 10 weeks after potting (WAP) and shoot dry weight were recorded at trail conclusion (10 WAP).

All data were analyzed using mixed model analysis of variance with statistical software (JMP® Pro ver. 14, SAS Institute, Cary, NC) with replication as a random factor and all other factors as fixed. Data were inspected to ensure the assumptions of ANOVA were met and then post hoc means separation was performed using Tukey's Honest Significant Differences test at a 0.05 significance level.

RESULTS AND DISCUSSION

Effect of substrate composition on growth of ligustrum. At 2 months after transplanting, growth index measurements showed plants were smaller in stratified substrates (1.3:S:7.5, 1.9:S:5, and 1.9:S:7.5) compared to the incorporated CRF substrates of 0.95:TO and 1.3:TO (Table 1). By 4 MAT, some treatment differences were observed, but all stratified treatments had growth indices similar to the 0.95:TO treatment. At 6 MAT, all stratified substrates had growth indices similar to standard incorporated treatments, indicating that while growth was initially reduced in stratified substrates, likely due to the unfertilized layer in the top 5 to 7.5 cm, the reduced growth index was only transient (Fig. 1). However, dry wt. data collected at trial conclusion revealed that while all stratified substrates had similar root and shoot biomass compared with the 0.95:TO treatment, 1.9:S:5 and 1.9:S:7.5 had lower shoot and root wt. compared with 1:3:TO.

Table 1. Effect of substrate composition on growth index and biomass of container-grown ligustrum (*Ligustrum japonicum*).

	Growth index (cm)[a]			Biomass[b]	
Substrate[c]	2MAT	4MAT	6MAT	Shoot wt (g)	Root wt (g)
0.95:TO	16.7 a	27.6 abc	45.8 ab	58.7 abc	25.0 abcd
1.3:S:5	17.5 a	28.8 ab	45.0 ab	61.3 ab	28.2 a
1.3:S:7.5	13.7 b	24.1 bc	42.8 ab	47.7 bc	20.3 bcd
1.9:S:5	13.8 b	22.8 c	40.8 b	42.4 c	19.2 d
1.9:S:7.5	13.6 b	23.8 bc	43.1 ab	44.6 bc	20.0 cd
1.3:TO	17.3 a	30.1 a	49.2 ab	60.8 ab	26.8 ab
1.9:TO	16.1 ab	28.7 abc	45.6 ab	58.6 abc	26.0 abc
1.9:S:5: N/F	17.8 a	28.3 abc	50.8 a	68.8 a	27.8 a

[a]Growth index was determined by calculating [(height+ width at widest point + perpendicular width) ÷ 3] from 2 to 6 months after transplanting (MAT). First experimental run was initiated on (Nov. 20, 2019) and second on (Feb. 2, 2020).

[b]Shoot and root dry wt. taken at trial conclusion at 6 months after transplanting (MAT).

[c]Substrate consisted of incorporated substrates with 0.95, 1.3 or 1.9 cm pine bark (PB) throughout the container with controlled release fertilizer (Osmocote® 17-5-11, 8 to 9 mo.) incorporated at 35 g pot-1 throughout (TO) (0.95:TO, 1.3:TO, and 1.9:TO) and stratified substrate treatments consisted of having the top substrate (1.3 or 1.9 cm pine bark) applied at depths of 5 or 7.5 cm, resulting in four stratified (S) substrate treatments (1.3:S:5, 1.3:S:7.5, 1.9:S:5, and 1.9:S:7.5) with the bottom substrate consisting of 0.95 cm bark with CRF incorporated. 1.9:S:5: N/F consisted of 1.9 cm bark with fines removed, applied at a 5 cm depth with 0.95 cm bark as the bottom substrate with CRF incorporated.

[d]Means followed by the same letter within a column are not significantly different according to Tukey's HSD test $\alpha = 0.05$.

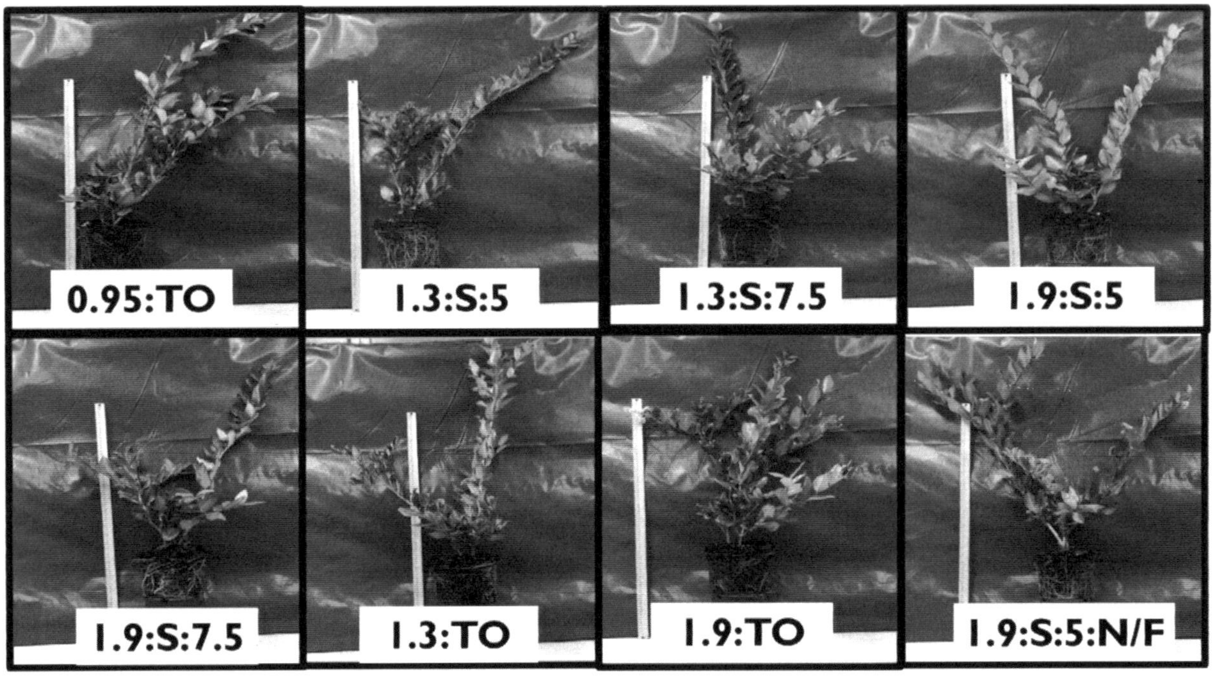

Figure 1. Growth of ligustrum (*Ligustrum japonicum*) at 6 months after transplanting. 0.95:TO = 0.95 cm pine bark (PB) incorporated with CRF throughout the container, 1.3:S:5 = 1.3 cm PB as top substrate applied at a 5 cm depth with 0.95 cm bark as the bottom substrate with CRF incorporated, 1.3:S:7.5 =1.3 cm PB as top substrate applied at a 7.5 cm depth with 0.95 cm bark as the bottom substrate with CRF incorporated, 1.9:S:5 =1.9 cm PB as top substrate applied at a 5 cm depth with 0.95 cm bark as the bottom substrate with CRF incorporated, 1.9:S:7.5 = 1.9 cm PB as top substrate applied at a 7.5 cm depth with 0.95 cm bark as the bottom substrate with CRF incorporated, 1.3:TO = 1.3 cm pine bark (PB) incorporated with CRF throughout the container, 1.9:TO = 1.9 cm pine bark (PB) incorporated with CRF throughout the container, 1.9:S:5:N/F = 1.9 cm bark with fines removed, applied at a 5 cm depth with 0.95 cm bark as the bottom substrate with CRF incorporated.

Interestingly, the 1.9:S:5:N/F treatment which had the lowest water holding capacity in the top 5 cm (data not shown) had shoot and root dry wt. similar to or greater than all incorporated treatments, even though no fertilizer was incorporated into the top 5 cm of the substrate.

Effect of substrate composition on bittercress (Cardamine flexuosa) germination and biomass. At 4 WAP, bittercress germination was highest in the 0.95:TO treatment followed by the 1.9:TO treatment which was similar while germination was lowest in the stratified substrate treatments (Table 2). The lowest germination was observed in the stratified substrate of 1.9:S:5: N/F (Table 2). By 8 WAP, fewer differences were observed in germination, but germination was significantly reduced in stratified treatments including 1.3:S:5, 1.9:S:5, and 1.9:S:5: N/F compared with 0.95:TO (Fig. 2).

Table 2. Effect of substrate composition on bittercress (*Cardamine flexuosa*) germination and biomass.

	Bittercress		
	Germination count pot^{-1}[a]		
Substrate[c]	4 WAP	9 WAP	Biomass[b]
0.95:TO	13.2 a[d]	14.0 a	7.6 a
1.3:S:5	6.9 cd	8.7 bc	1.2 b
1.3:S:7.5	6.7 cd	10.1 abc	0.5 b
1.9:S:5	6.9 cd	8.1 bc	1.1 b
1.9:S:7.5	7.9 bc	10.4 ab	0.5 b
1.3:TO	8.6 bc	11.9 ab	5.9 a
1.9:TO	11.3 ab	11.8 ab	7.7 a
1.9:S:5: N/F	3.4 d	6.2 c	0.2 b

[a]Germination count was assessed by surface sowing 25 seeds of bittercress (*Cardamine flexuosa*) to each pot and counting germinated seedling at 4 weeks and 9 weeks after potting (WAP). First experimental run was initiated on (Nov. 20, 2019) and second on (Feb. 2, 2020).

[b]Shoot dry wt. was taken at trial conclusion at 10 weeks after potting.

[c]Substrate consisted of incorporated substrates with 0.95, 1.3 or 1.9 cm pine bark (PB) throughout the container with controlled release fertilizer (Osmocote® 17-5-11, 8 to 9 mo.) incorporated at 35 g pot-1 throughout (TO) (0.95:TO, 1.3:TO, and 1.9:TO) and stratified substrate treatments consisted of having the top substrate (1.3 or 1.9 cm pine bark) applied at depths of 5 or 7.5 cm, resulting in four stratified (S) substrate treatments (1.3:S:5, 1.3:S:7.5, 1.9:S:5, and 1.9:S:7.5) with the bottom substrate consisting of 0.95 cm bark with CRF incorporated. 1.9:S:5: N/F consisted of 1.9 cm bark with fines removed, applied at a 5 cm depth with 0.95 cm bark as the bottom substrate with CRF incorporated.

[d]Means followed by the same letter within a column are not significantly different according to Tukey's HSD test $\alpha = 0.05$.

Bittercress dry wt. taken at trial conclusion revealed that all stratified treatments had significant lower dry wt. compared with incorporated treatments, with stratification and strategic fertilizer placement resulting in decreases ranging from 80 to 97% in bittercress biomass. The top layer of stratified treatments had no fertilizer in top 5 or 7.5 cm of the substrate, hence the top layer lacked nutrients for bittercress growth. Similar results were reported by Saha et.al (2019), where 90% reduction in eclipta growth was observed when fertilizer was subdressed resulting in a similar effect where no nutrients were contained on the surface of the substrate.

Figure 2. Bittercress (*Cardamine flexuosa*) at 10 weeks after potting. 0.95:TO = 0.95 cm pine bark (PB) incorporated with CRF throughout the container, 1.3:S:5 = 1.3 cm PB as top substrate applied at a 5 cm depth with 0.95 cm bark as the bottom substrate with CRF incorporated, 1.3:S:7.5 =1.3 cm PB as top substrate applied at a 7.5 cm depth with 0.95 cm bark as the bottom substrate with CRF incorporated, 1.9:S:5 =1.9 cm PB as top substrate applied at a 5 cm depth with 0.95 cm bark as the bottom substrate with CRF incorporated, 1.9:S:7.5 = 1.9 cm PB as top substrate applied at a 7.5 cm depth with 0.95 cm bark as the bottom substrate with CRF incorporated, 1.3:TO = 1.3 cm pine bark (PB) incorporated with CRF throughout the container, 1.9:TO = 1.9 cm pine bark (PB) incorporated with CRF throughout the container, 1.9:S:5:N/F = 1.9 cm bark with fines removed, applied at a 5 cm depth with 0.95 cm bark as the bottom substrate with CRF incorporated.

Stewart et al (2017) also showed that depths of 2.5 cm are efficient to control spotted spurge, large crabgrass (*Digitaria sanguinalis*), bittercress and liverwort (*Marchantia polymorpha*) growth. While the growth of bittercress decreased in stratified substrates, it could be due to the strategic placement of fertilizer and more research is needed to quantify the effect of stratification on weed germination and growth.

Results from this study showed that substrate stratification and strategic fertilizer placement may result in early growth reductions to ligustrum, but no differences were observed between stratified substrates and an industry standard 0.95:TO substrate by the time plants reached marketable size or were ready to be transplanted into larger containers. The growth of bittercress substantially decreased in the stratified substrates indicating that stratified substrate could potentially be used as part of an overall weed management strategy. As these methods of substrate stratification and strategic fertilizer placement have not been evaluated widely, additional research is currently being conducted to determine the effects of these strategies on the growth of other common ornamental species and problematic weeds in container nurseries, as well as the effects of these methods on irrigation and fertilization efficiency.

Literature Cited

Berchielli-Robertson, D.L., Gilliam, C.H., and Fare, D.C. (1990). Competitive effects of weeds on the growth of container grown plants. HortScience *25*:77–79.

Case, L.T., Mathers, H.M., and Senesac, A.F. (2005). A review of weed control practices in container nurseries. HortTech. *15*:535-539.

Fields, J.B., Owen, J.S., and Altland, J. (2020). Substrate stratification: A media management strategy for increased resource efficiency. Proc. Amer. Soc. Hort. Sci. South. Reg. (In press).

Gilliam, C.H., Fare, D.C., and Beasley, A. (1992). Non-target herbicide losses from application of granular Ronstar to container nurseries. J. Environ. Hort. *10*:175–176.

Harper, J., and Benton, R. (1966). The behavior of seeds in soil. II. The germination of seeds on the surface of a water supplying substrate. J. Ecol. *54*: 151-166.

Keddy, P.A., and Constabel, P. (1986). Germination of ten shoreline plants in relation to seed size, soil particle size and water level: an experimental study. J. Ecol. *74*:133-141.

Owen, J.S., and Altland, J.E. (2008). Container height and douglas fir bark texture affect substrate physical properties. HortScience *43*:505-508.

Saha, D., Marble, C., Torres, N., and Chandler, A. (2019). Fertilizer placement affects growth and reproduction of three common weed species in pine bark-based soilless nursery substrates. Weed Sci. *67*:682-688.

Stewart, C., Marble, C., Jackson, B., Pearson, B., and Wilson, C. (2017). Impact of container nursery production practices on weed growth and herbicide performance. HortScience *52*:1593–2000.

Stewart, C., Marble, C., Jackson, B., Pearson, B., and Wilson, C. (2019). Effects of three fertilization methods on weed growth and herbicide performance in soilless nursery substrates. J. Environ. Hort. *36*:133-139.

Wada, S. (2005). Nursery container weeds response to modification of substrate pH, substrate particle size and applied nitrogen form. Oregon State Univ., Corvallis, MS Thesis, 72–76.

Combined Proceedings IPPS 70:38-46. 2020.

Where Did the Water Go?

Brian A. Schulker[a1], Brian E. Jackson[1], and William C. Fonteno[1]

[1]Department of Horticulture, North Carolina State University, Raleigh, NC 27695

baschulk@ncsu.edu

[a]Third Place- Charlie Parkerson Graduate Student Research Paper Competition

Keywords: Capillarity, subirrigation, hydrophobicity, substrate, macropore, micropore

Summary

Particles in a substrate create a network of pathways for water to move through, with size and shape determining the efficacy of these channels. Reduced particle size diversity can lead to excessive leachate, poor substrate hydration, and an inefficient irrigation practice. This research was designed to examine the water capture characteristics of peat, coir, and pine bark using three initial moisture contents (MC) of 67%, 50%, and 33% (by weight) through subirrigation under three time-interval pulse irrigation regimens. The objective was to determine the impact of differing irrigation event durations (5, 10, 20) over a 60-min total period of time, water depth, and initial moisture on the initial water capture rate of these three substrates. Initial capture rate (ICR) was influenced by MC, irrigation water depth, and inherent substrate characteristics (hydrophobicity / hydrophilicity). Initial moisture content had the greatest impact on peat, regardless of water depth or pulsing time. Lower moisture conditions increased the hydrophobic characteristics of peat, lessening the amount of water it was able to capture in the first irrigation event with the ICR of peat never reaching 1 mL/min at 33% MC. Pine bark had a 2 mL/min decrease in initial capture rate across 67, 50, and 33% MCs, while coir's hydrophilic nature reduced any moisture content affects. At 50% MC or less, coir had the highest capture rate across all substrates, pulsing durations, and water depths. Water depth was found to increase capture 2-4 mL/min across all substrates (aside from 33% MC peat). While pulsing time produced variable results, with an increase in pulsing time not always equaling the added volume of water from the 5 min treatment. Ultimately, these three substrates portrayed benefits to irrigation capture that further research is needed to understand. Pine bark captured more water under low moisture equal to or better than 50% and 67% MC while coir and peat exhibited higher water retention abilities (peat at higher MCs). Engineering substrates to combine or enhance these characteristics could allow growers to decrease irrigation rates and frequencies while still producing healthy, viable crops. It is believed to be feasible to select substrate components (or types) to fit the irrigation delivery method and container type of a grower to achieve maximum irrigation efficiency for different crops.

INTRODUCTION

Water use efficiency of horticultural soilless substrates represents one of the largest variables in container plant production. With nearly 21,500 acres of land devoted to greenhouse operations in the U.S., representing a 148% increase since 1998, growers specializing in container plant production need to understand how irrigation technique can impact the water use efficiency of soilless substrates (USDA, 2014). As water quality, conservation, and scarcity concerns increase, as well as operational costs, growers must adopt new strategies to maintain the sustainable use of water to confront water-climate policy (González et al., 1992; Deccache et al., 2014; Egea et al., 2017; Montesano et al., 2018). Understanding the components that make up a substrate allows a better understanding of substrate properties and their management and further increases the ability to structure substrate tendencies to production practices. Components are classified as the individual materials (peat, coir, pine bark, perlite, wood, etc.) that, when mixed together, make up the substrate.

Whether ebb-and-flow, flood-floor, capillary mat or other systems, subirrigation is a popular technique in container plant production with the ability to control the application of water, further increasing the operation efficiency (Dole et al., 1994; Uva and Weiler, 2001). The economic benefits of a subirrigation system can be a lower labor requirement (compared to traditional overhead systems) and an even application of water, leading to a more uniform crop (Elliott, 1990; Uva et al., 1998). Subirrigation systems have the ability to reduce application runoff (Klock-Moore and Broschat, 1999) further reducing water and fertilizer costs, which are a few key points in the use of this irrigation practice. Compared to overhead and surface irrigation systems, subirrigation was found to consistently reduce overall water use due to the recollection and reuse of water in the system (Davis et al., 2008, 2011; Dumroese et al., 2007; Elliott, 1990). The objective of this experiment was to understand the effect of pulsing irrigation techniques and initial moisture content on the initial water capture of peat, coir, and pine bark substrates using ebb and flood subirrigation.

MATERIALS AND METHODS

Substrate components being tested were sphagnum peatmoss (Premier Pro-Moss, Quakertown, PA), coconut coir (Densu Coir, Ontario, Canada), and pine bark (Pacific Organics, NC). Peat was removed from the compressed bale, water was added and peat was agitated by hand to allow proper water absorption. Moisture contents (MCs) were then tested in order to bring the MC up to 70% before being dried down to MCs of 67%, 50%, and 33%. Compressed bricks of coconut coir were hydrated by adding 14 L of water by hand, until the compressed brick was completely broken apart before testing moisture levels to ensure an initial MC of 70%. Four-month aged loblolly (*Pinus taeda* L.) pine bark was weighed, moisture levels tested, and further hydrated to a moisture content of 70%. Cylinders were then packed by weight, keeping all 4 reps of each substrate moisture content within 5% of each other and then packing them down to a pre-measured 10 cm of height to ensure similar bulk densities.

The equipment used follows the same procedure as Schulker (2020), and consisted of a transparent cylinder, 5 cm i.d. x 15 cm·h^{-1}, with a mesh screen (mesh size 18 x 16; New York Wire, York, PA), attached to one end, using rubber pressure plate rings (Fig. 1B). The subirrigation method used to simulate capillary rise uses an Ebb and Flood irrigation unit (Hawthorn Hydroponics, Vancouver WA) 2ft wide by 4ft in length (Fig. 1A).

Pulsing in this context means time of exposure to water based on a total time of 60 minutes. The hydration events tested were 5min exposure - 12 events, 10min exposure – 6 events, and 20min exposure – 3 events. All of which were tested at water heights of 2mm, 20mm, and 35mm above the sample base. Once placed on the mesh screen, the unit was filled with water. Water was allowed to fill until water poured into the copper piping fitted to the desired water level. At that time, water flow input equaled output, allowing constant flow of water without a change in water level. The substrates were held at a constant water level for the allotted event time (between 5 min and 20 min), once finished, water was drained from the unit for one timed minute before each cylinder was weighed. The difference between final and initial weights was the amount of water captured by the substrate during hydration. This procedure was repeated based on specified time-allotted events (12, 6, and 3 irrigation event(s)).

Initial capture rate. Initial capture rate (ICR) was calculated using a version of the flow rate formula to account for variables in this experiment, the equation was written as

$$\mathrm{ICR} = \frac{C_i - C_p}{t} \qquad (1)$$

where ICR is the amount of water captured by the substrate after the first irrigation per unit time (in mL/min), C_i (initial capture) is the weight (g) of the substrate after the present irrigation event (minus the weight of the cylinder), C_p (previous capture) is equal to the pack weight of the cylinder (minus the weight of the cylinder), t is the amount of time per irrigation (minutes).

Figure 1. Ebb and flood subirrigation system. A) Fully constructed system complete with packed substrate cylinders during experimentation. B) Close up of 2mm water level during irrigation event.

RESULTS

Initial capture rate (ICR) was calculated for each pulsing time, water level, and MC of coir by the equation (1) and recorded in Table 1. Based on the formula used, the CR falls as pulsing time increases, as that would increase the t-value in the denominator of the equation. However, that does not mean the amount of water captured is any less, the water has more time to be absorbed by the substrate.

Table 1. Initial capture rate (ICR) (in mL/min) at three initial moisture contents (MC), three irrigation water levels, and at three irrigation pulse durations per water level.

ICR[z]	Peat			Coir			Bark		
	33%IM	50%IM	67%IM	33%IM	50%IM	67%IM	33%IM	50%IM	67%IM
2mm[x]									
5 min	0.29 c[y]	0.71 d	3.72 d	9.66 b	9.97 c	8.77 c	6.73 bc	4.91 c	3.99 c
10 min	0.06 d	0.72 d	2.40 de	4.59 c	4.75 e	4.50 d	3.06 d	2.48 d	2.00 d
20 min	0.03 d	0.64 d	1.48 e	2.64 d	2.84 f	2.60 e	1.71 e	1.40 e	1.03 e
20mm									
5 min	0.31 b	3.25 b	7.34 b	10.83 b	12.74 b	12.29 b	8.36 b	8.62 b	7.16 b
10 min	0.18 bc	2.27 bc	4.29 cd	5.94 c	6.76 d	6.47 cd	4.24 c	4.50 c	3.87 c
20 min	0.13 c	1.64 c	2.41 de	3.48 cd	3.66 e	3.42 d	2.47 de	2.35 d	2.04 d
35mm									
5 min	0.61 a	6.04 a	9.49 a	14.27 a	15.43 a	14.02 a	10.61 a	11.26 a	9.01 a
10 min	0.34 b	3.83 b	5.34 c	7.53 bc	8.47 c	7.29 c	5.98 c	5.65 c	4.70 c
20 min	0.30 b	2.52 bc	2.97 d	4.09 c	4.34 e	3.89 d	2.74 de	2.98 d	2.45 d

[z]ICR = the amount of water (in mL per min) that each substrate is able to capture after one irrigation per unit time. [y]Statistics using Tukey's honestly significant difference with alpha = 0.05 are given down individual columns at a given initial moisture content. [x]Water depth during irrigation event expressed in millimeters.

Coir. Initial CR was directly affected by water level and pulsing duration. Based on the formula used, it is understandable that the ICR decreases as pulsing time increases (Fig. 2), as that would increase the t-value in the denominator of the equation. However, that does not mean the amount of water captured is any less, the water simply has more time to be absorbed by the substrate. At 20mm, there is flooding (of the cylinder) involved, increasing the CR compared to 2mm which is based solely on capillary movement of water by the substrate. For coir, there is an incremental increase in water captured based on MC and water depth, with 50% MC and 5min time interval representing the highest ICR. Even at 2mm, coir is able to capture water at nearly the same rate as 20mm, exhibiting the hydrophilic nature of the material.

Peat. Initial CR was calculated for peat using the same formula that was used for coir (Table 1). Based on the effects MC had on the water capture of peat, the values for ICR at 33% IMC were lower than all other substrates tested. Increasing time (further increasing the value for t in the equation) did not have a major effect in the values at 33% MC. Increasing the moisture to 50% MC, in relation to 33% IMC, impacts that CR of peat. With an increase in ICR as much as 6 mL/min at a 20mm water depth. Peat, under lower moisture conditions was unable to capture water in the same manner as coir (Fig. 2). Even at 35mm, the greatest capture rate for low moisture peat did not crest 1 mL/min.

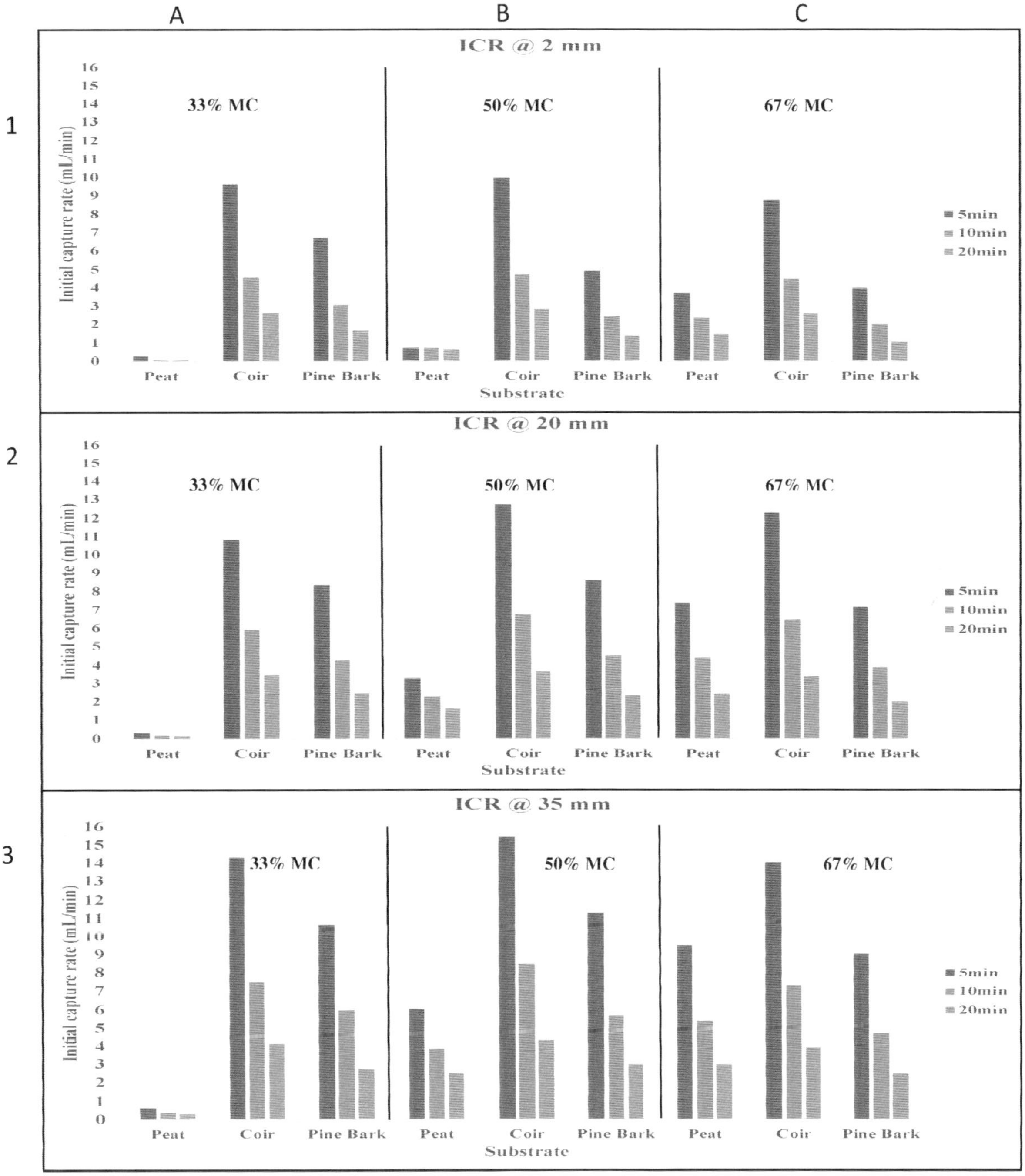

Figure 2. Substrate initial capture rate (ICR) for peat, coir, and pine bark over three initial moisture contents (MC) of 33%, 50%, and 67%, three water levels of 2mm, 20mm, and 35mm, and three pulse durations of 5min (blue), 10min (orange), and 20min (gray).

Pine Bark. Initial CR for each treatment showed just how consistent pine bark was. Reversing the equation, by re-multiplying by the number of events away from 5min, showed that pine bark captured water at nearly the same rate across all pulsing times within the same MC. Compared to both peat and coir, pine bark captured the majority of water within the first 5min irrigation pulse (Fig. 2). the ICR for pine bark did not increase with MC, in most cases it actually decreased. With 33% MC representing the highest ICR for most pine bark treatments

DISCUSSION

From the values shown in Table 1, it appears that initial MC prior to the first irrigation event and depth had the greatest effect on the ICR of peat, coir, and pine bark.

Across all initial MCs and water depths, coir was able to capture and retain the most water comparatively. However, MC played a role in determining just how fast it was able to do so. At 5min and 2mm, there was very little difference in the ICR based on MC, and that holds true for all pulsing times at that water level. Abad et al. (2005) characterizes coir as having a sponge-like ability to soak up water and be able to retain it within the pores of the substrates, and this is evident in table 1. As water level increases, the ICR increases. Showing that the increase in irrigation water depth truly plays the biggest role in the amount of water coir captures in the first irrigation event. Increasing by 5 mL/min from 2mm to 35mm.

It is evident from the data in Table 1, that initial MC had the greatest impact on the ability of peat to capture water, regardless of pulsing time or water depth. As moisture levels increased in peat, the substrates ability to capture water increased, in a nearly linear fashion. Peat is known to have hydrophobic characteristics which could come from inherent characteristics of a substrate at lower moisture levels (Michel et al., 2001) or from material drying processes in the production of these substrates. At 33% MC, peat exhibited difficulty in capturing any water whether it was a 5min pulse at 2mm or a 20min pulse at 35mm. The hydrophobic tendencies truly hindered the ability of peat to rewet, taking more water and time to wet the substrate. The main result shown is that the ICR of peat is nearly 10 mL/min less under lower moisture conditions, representing the largest difference of the three substrates.

Pine bark was comparably unaffected by pulse time and observed an increase in ICR as water depth increased from 2mm to 35mm. Similar to both peat and coir, the higher the initial MC and water depth, the greater the capture. Generally speaking, pine bark is known to have larger particle sizes than both peat and coir. The larger pore sizes created by these larger particles tend to have difficulty holding water after saturation (Drzal et al., 1999). However, these larger pore sizes aided in pine bark to capture water at lower MCs, with 33% moisture exhibiting the highest ICR at 2mm, and within 1 mL/min for 50% MC and 67% MC (Table 1).

CONCLUSION

Initial CR was designed to be able to understand the first irrigation characteristics of these substrates, and how different variables such as water depth and pulsing time can affect the amount of water captured by the substrate. The results of this study showed that initial MC had the biggest role to play in the ability of materials to take up water, with peat showing the greatest difference in water capture. Coconut coir captured and retained water in a sponge-like manner regardless of treatment while pine bark showed little variation based on MC. Overall, these three substrates represented differing abilities in water capture through water level, initial MC, and pulsing time. With each of the three substrates being tested representing different particle size distributions, if we could manipulate the particle size fractions and/or percentages in these substrates it could fundamentally change the initial capture rate, making irrigations more efficient while conserving more water. In doing so, continued research on engineering and formulating substrates with the goal of increasing the water capture efficiency of substrates is of potential great significance to the future of precision growing of plants in container systems. For example, blending different substrate components (at varying percentages) with different water capturing abilities to enable maximum container substrate hydration with the fewest irrigations as possible could reduce the inconsistences of container crop irrigation scheduling and practices. It is believed to be feasible to select substrate components/types to fit the container and irrigation delivery method of a grower to achieve maximum irrigation efficiency for different crops.

Literature Cited

Abad, M., Frones, F., Carrion, C., and Noguero. V. (2005). Physical properties of various coconut coir dusts compared to peat. HortScience *40*:2138– 2144.

Daccache, A., Ciurana, J.S., Rodriguez Diaz, J.A., Knox, J.W. (2014). Water and energy footprint of irrigated agriculture in the Mediterranean region. Environ. Res. Lett. *9*:1–12.

Davis, A.S., Pinto, J.R., and Jacobs, D.F. (2011). Early field performance of *Acacia koa* seedlings grown under subirrigation and overhead irrigation. Native Plants J. *12*:94–99.

Dole, J.M., Cole, J.C., and von Broembsen, S.L. (1994.) Growth of poinsettias, nutrient leaching, and water-use efficiency respond to irrigation methods. HortScience *29*:858–864.

Drzal, M.S., Fonteno, W.C., and Cassel, D.K. (1999). Pore fraction analysis: A new tool for substrate testing. Acta Hortic. *481*:43–54.

Dumroese, R.K., Jacobs, D.F., Davis, A.S., Pinto, J.R. and Landis, T.D. (2007). An Introduction to Subirrigation in Forest and Conservation Nurseries and Some Preliminary Results of Demonstrations. U.S. Dept. Agr., For. Serv. Proc. RMRSP-50:20–26.

Egea, G., Fernández, J.E., and Alcon, F.(2017). Financial assessment of adopting irrigation technology for plant-based regulated deficit irrigation scheduling in super high-density olive orchards. Agric. Water Manag. *187*: 47–56.

Elliott, G.C. (1990). Reduce water and fertilizer with ebb and flow. Greenhouse Grower *8*:70–73.

Klock-Moore, K.A. and Broschat, T.K. (1999). Differences in bedding plant growth and nitrate loss with a controlled–release fertilizer and two irrigation systems. HortTech. *9*:206–209.

Michel, J.C., Riviere, L.M., and Bellon Fontaine, M.N. (2001). Measurement of wettability of organic materials in relation to water content by the capillary rise method. European J. Soil. Sci. *52*:459–467.

Montesano, F.F., Van Iersel, M.W., Boari, F., Cantore, V., D'Amato, G., and Parente, A. (2018). Sensor-based irrigation management of soilless basil using a new smart irrigation system: Effects of set-point on plant physiological responses and crop performance. Agric. Water Manag. *203*:20–29.

Schulker, B. A. (2020). Substrate Conditioning Influences Water Capture Efficiency of Common Greenhouse Substrate Components. M.S. Thesis, NC State University, Raleigh.

U. S. Department of Agriculture (2007). Nursery crops 2006 summary. In National Agricultural Statistics Serv. (Ed.), Washington, DC., p. 63 pp.

Uva, W.L. and Weiler. T.C. (2001). Economic analysis of adopting zero runoff subirrigation systems in greenhouse operations in the northeast and north central United States. HortScience *36*:167–173.

Combined Proceedings IPPS 70:47-51. 2020.

Evaluation of Honey as a Rooting Aid for the Propagation of Rosa 'Red Cascade'

Anthony T. Bowden, Patricia R. Knight, Christine E.H. Coker, Jenny B. Ryals, Scott A. Langlois, Shaun R. Broderick, Eugene K. Blythe, Hamidou F. Sakhanokho, and Ebrahiem M. Babiker

Coastal Research and Extension Center, South Mississippi Branch Experiment Station, P.O. Box 193, Poplarville, MS 39470 USA

ab1001@msstate.edu

Keywords: Auxin, cutting propagation, indole-3-butyric acid (IBA)

Summary

Previous research has shown that honey may prove beneficial to the plant propagation process. The objective of this research was to evaluate whether addition of honey to water-soluble auxin solutions increased root growth and uniformity compared to auxin solutions without honey on medial cuttings of *Rosa* 'Red Cascade'. The 4×5 factorial experiment consisted of four honey types (none, general multiflora, Manuka, or locally sourced honey), and five auxin levels (0, 250, 500, 750, or 1,000 ppm indole-3-butyric acid (IBA). Utilization of honey or auxin during propagation of 'Red Cascade' miniature rose did not increase percent rooting, number of roots, root length, shoot height, or root quality rating for cuttings. Additionally, honey type and auxin rate had no effect on net photosynthetic rate and stomatal conductance. Further research is being conducted with other woody ornamental plant species that vary in rooting difficulty to determine if addition of various honey types to water soluble IBA solutions enhances rooting responses.

INTRODUCTION

Honey has a long history as a wound treatment, dating back to the writings of the Egyptians 5,000 years ago. In these earliest writings, honey was reportedly made into an ointment to treat skin and eye diseases - as well as applied as a dressing for burns and wounds (Israili, 2003). In the nursery industry, sugars (carbohydrates) are known to positively impact rooting of cuttings and are frequently used in tissue culture as an energy source for micro-cuttings (Davies et al., 2018). Clinical studies have confirmed the broad-spectrum antimicrobial properties of honey which are theorized to be due to naturally low pH, osmotic effect, high sugar concentration, and presence of bacteriostatic and bactericidal factors (Israili, 2003). Antibacterial properties are attributed to the super-saturated solution of sugar (Molan, 1992). A typical batch of honey has a 15-21% moisture content and a solid fraction containing a ratio of monosaccharides (glucose and fructose) which leaves very little free water available for growth of micro-organisms (Molan, 2001). Antimicrobial properties of manuka honey were determined by testing and the results used to calculate the unique Manuka factor (UMF) which ranges in potency from 1 -70^{+} (Whalley, 2009).

Whalley (2009) of Taupo Native Plant Nursery in New Zealand conducted an on-nursery trial using honey as a stand-alone rooting hormone when their primary rooting hormone powder was discontinued. Whalley trialed Manuka honey (UMF 15+), a multiflora honey purchased from the supermarket, a commercially available root-promoting compound, and a nontreated control. These treatments were applied to cuttings of six New Zealand natives: *Brachyglottis* 'Sunshine', *Coprosma acerosa, Coprosma* x *kirkii* 'Kirkii', *Griselinia littoralis* 'Broadway Mint', *Myoporum laetum*, and *Olearia virgata* var. *lineata*. These plants were selected for various characteristics including ease of rooting (Whalley, 2009). Both varieties of honey were used to prepare solutions containing honey and hot water (1:2 v:v) and solutions were refrigerated for 24-h before use (Whalley, 2009). Cuttings were placed into solutions for 30 minutes before sticking and placed onto a mist bed with bottom heat (Whalley, 2009). Cuttings treated with solutions containing multiflora honey had the fewest unrooted cuttings across all four treatments and a high number of cuttings with a good (4+) and average (3 or less) root rating. Cuttings treated with Manuka honey had the lowest number of roots with a good (4+) rating and a higher number of unrooted cuttings among all four tested treatments (Whalley, 2009). Previous research has shown that honey may prove beneficial to the plant propagation process; however, further research is needed to quantify if honey proves beneficial in different scenarios of cutting treatment times. The objective of this research was to evaluate whether addition of honey to water-soluble auxin solutions increased root growth and uniformity compared to auxin solutions without honey.

MATERIALS AND METHODS

Multi-node medial cuttings of *Rosa* 'Red Cascade' were harvested from containerized stock plants, trimmed to 2.5-in (6.4 cm). in length and stuck to a depth of 0.5-in. (1.3 cm) on 22 May 2020. Red Cascade rose was chosen as the model plant for preliminary studies since previous research has shown that it can be rooted successfully without using auxin although using an auxin-containing compound can result in an increased rooting response (Blythe et al., 2003). Propagation medium was 100% pine bark placed into 3.5-in. (8.3 cm) square production pots (T.O. Plastics, Inc., Clearwater, MN).

Cuttings were placed under intermittent mist applied for 6-sec/10-min during daylight hours. Treatments consisted of four honey types (none, general multiflora, Manuka, or locally sourced), and five auxin levels [0, 250, 500, 750, or 1,000 ppm indole-3-butyric acid (IBA) (Hortus Water Soluble Salts; Phytotronics Inc., Earth City, MO)]. Water soluble IBA solutions were created using deionized water. Honey treatments consisted of a 2:1 solution created by dissolving honey in either deionized water (when auxin level equaled 0 ppm) or the IBA solutions. Once the solutions were made, cuttings were treated with a 1-sec basal quick-dip in one of the twenty solutions before being stuck into production flats.

A completely randomized design with a 4×5 factorial treatment arrangement was utilized with 15 cuttings per treatment. Data collected after 42 days included rooting percentage, shoot height, total root number, average root length (three longest roots), and root quality (1-5, with 1=no roots and 5= ≥ 10 roots).

Additionally, net photosynthetic rate (A) and stomatal conductance (gs_W) values were sampled from five cuttings per treatment, for a total of 100 cuttings, between the hours of 7:30 A.M. and 11:30 A.M., using the LiCOR® 6800 Portable Photosynthesis System (LI-COR Biosciences; Lincoln, NE). Data were analyzed using linear mixed models and generalized linear mixed models with the GLIMMIX procedure of SAS (ver. 9.4; SAS Institute Inc., Cary, N.C.)

RESULTS AND DISCUSSION

There was no interaction between honey type and auxin rate therefore only main effects are presented (Table 1). Utilization of honey or auxin during propagation of 'Red Cascade' miniature rose did not increase percent rooting, number of roots, average length of three longest roots, shoot height, or root quality rating for cuttings (Table 1). Additionally, the interaction between honey type and auxin rate was non-significant for net photosynthetic rate and stomatal conductance (Table 2). Use of honey or auxin during the propagation process does not increase the rate of gas exchange during the rooting process and, as a result, the assimilation of CO_2 into dry matter is unaffected. The data taken from the LI-COR platform further confirms that our tested rooting parameters as well as subsequent shoot growth is neither helped nor hindered by the addition of honey to water-soluble IBA solutions.

Table 1: Results of three different honey sources on rooting percentage, root number, average root length, root quality, and growth of medial stem cuttings of a miniature climbing rose (*Rosa* 'Red Cascade').

	Rooting (%)	Roots (no.)	Avg. Length of three longest roots (cm)	Shoot Height (cm)	Root Quality Rating[z]
Honey Type:					
No Honey	83.3	1.4ab	9.7a	3.4a	2.9a
Local Honey	100	1.4ab	8.6a	2.9a	2.8a
Manuka Honey	100	1.3b	9.5a	3.8a	2.6a
Multiflora Honey	100	1.6a	8.8a	3.4a	2.9a
Auxin Rate:					
0 ppm IBA	83.3	1.5a	8.6a	3.4a	2.8a
250 ppm IBA	100	1.5a	7.8a	2.6a	2.6a
500 ppm IBA	100	1.4a	9.6a	3.5a	2.8a
750 ppm IBA	93.3	1.4a	9.9a	3.6a	2.7a
1,000 ppm IBA	100	1.5a	9.8a	3.8a	2.8a
Significance[x]:					
Honey Type	NS	NS	NS	NS	NS
Auxin Rate	NS	NS	NS	NS	NS
Honey Type × Auxin Rate	NS	NS	NS	NS	NS

[z]Root Quality (1-5, with 1 = no roots and 5 = ≥ 10 roots)
[y]Means within a column followed by the same letter were not different at $\alpha = 0.05$ or 0.10.
[x] Significant at the $P \leq 0.1$ (*) or 0.05 (**) level. NS= Not significant

Our results differ from Whalley (Whalley, 2009). In their experiments, using a Manuka or a multiflora honey enhanced rooting, including a healthier or higher quality root system, when compared to a commercial root-promoting compound. Our results showed that honey type did not impact our tested rooting parameters. Our results differ from previous experiments on the effects of IBA rate on the rooting of Red Cascade miniature rose (Blythe et al., 2003). In their experiment, they found that rooting response of 'Red Cascade' rose increased with increasing auxin concentration. Our results showed that regardless of auxin concentration, our tested rooting parameters are not impacted by increasing auxin rate. Cuttings treated with 1,000 ppm IBA rooted similarly to those cuttings not receiving IBA at the time of treatment initiation.

Further research is currently ongoing to examine other woody ornamental plant species that vary in rooting difficulty to determine if addition of various honey types to water soluble IBA solutions enhances rooting responses. Cuttings that require longer propagation times may benefit from the addition of honey to auxin solutions as longer propagation time can allow for an extended period for soil pathogens to impact cutting health.

Table 2: Results of three different honey sources on assimilation rate and stomatal conductance values of medial stem cuttings of a miniature climbing rose (*Rosa* 'Red Cascade').

	Assimilation Rate (A) (μmol·m^{-2}·s^{-1})	Stomatal Conductance (gs_w) (mol·m^{-2}·s^{-1})
Honey Type:		
No Honey	7.1a^z	0.19a
Local Honey	8.7a	0.19a
Manuka Honey	7.4a	0.16a
Multiflora Honey	8.6a	0.18a
Auxin Rate:		
0 ppm IBA	8.4a	0.16a
250 ppm IBA	9.0a	0.18a
500 ppm IBA	8.4a	0.19a
750 ppm IBA	6.8a	0.15a
1,000 ppm IBA	7.2a	0.20a
Significance^x:		
Honey Type	NS	NS
Auxin Rate	NS	NS
Honey Type × Auxin Rate	NS	NS

^{z}Means within a column followed by the same letter were not different at $\alpha = 0.05$ or 0.10.

^{y} Significant at the $P \leq 0.1$(*) or 0.05 (**) level. NS= Not significant

Literature Cited

Blythe, E.K., Sibley, J.L., Tilt, K.M., and Ruter, J.M. (2003). Cutting propagation with auxin applied via a stabilized organic rooting substrate. Comb. Proc. Intl. Plant Prop. Soc. *53*:275-283.

Davies, F.T., Geneve, R.L., and Wilson, S.B. (2018). Hartmann and Kester's Plant Propagation: Principles and practices. 9th ed. Prentice Hall, Upper Saddle River, N.J.

Israili, Z. H. (2014). Antimicrobial properties of honey. Amer. J. Therapeutics. *21*:304-323.

Molan, P. (1992). The antibacterial activity of honey: 1. The nature of the antibacterial activity. Bee World. *73*:5-28.

Molan, P. (2001). Why honey is effective as a medicine: 2. The scientific explanation of its effects. Bee World. *82*:22-40.

Whalley, L. (2009). Sweet Roots: A trial using honey as a rooting hormone. Comb. Proc. Intl. Plant Prop. Soc. *59*:74-77.

Combined Proceedings IPPS 70:52-59. 2020.

Mulch Depth Effect on Rooting Stem Cuttings and Weed Control During Propagation

Isha Poudel and Anthony Witcher

Department of Agricultural and Environmental Sciences, Otis L. Floyd Nursery Research Center, Tennessee State University, 472 Cadillac Lane, McMinnville, TN, 37110 USA

ipoudel@my.tnstate.edu

Keywords: *Cardamine hirsuta, Digitaria sanguinalis, Hydrangea paniculata* 'Phantom', nursery crops, weed efficacy

Summary

Hand weeding is the most common method for controlling weeds in nursery crop propagation, but it is time-consuming and costly due to high labor costs. Pre-emergence herbicides are not labeled to be used in non-rooted cuttings, but mulches may be a viable alternative to hand weeding and herbicides. The objective of this study was to determine the effect of mulch type (coarse vermiculite, paper pellets, pine pellets and rice hulls) and mulch depth [1.3 and 2.5 cm (0.5 and 1 in.)] on rooting stem cuttings and weed control in propagation. Hydrangea softwood cuttings (*Hydrangea paniculata* 'Phantom') were used for the rooting study while seeds of two weed species [large crabgrass (*Digitaria sanguinalis*) and bittercress (*Cardamine hirsuta*)] were used for the weed control study. Rooting percentage for hydrangea was 100% for all treatments except rice hulls and paper pellets applied at 2.5 cm (1 in.) depth (90% and 95%, respectively). Root dry weight and total root length were similar to the control for all mulches. Root volume was lowest for paper pellets at 2.5 cm (1 in.) depth, but similar to the control for all other treatments. Crabgrass seedling counts were similar for all treatments compared to the non-treated control. Bittercress seedling count was lower than the non-treated control for pine pellets at both mulch depths. Pine pellets and paper pellets suppressed shoot fresh weight of crabgrass and bittercress at both mulch depths. We conclude that pine pellets and paper pellets were found to be effective in controlling weeds in propagation, but further research should focus on rooting safety of other crop species.

INTRODUCTION

Weeds are a major issue in nursery crop production and the problem is more severe in propagation due to the small container volume where weeds compete with the crop for nutrients, light, and water (Altland, 2005). Weeds can also harbor different type of insects and plant pathogens which damage crops (Cranston, 1994; Hobbs et al., 1995). Due to weed infestations, nursery growers can face economic losses of $7000 per acre (Mathers, 2003). Hand weeding is the most common method of weed control in propagation, but hand weeding can suppress growth of cuttings through mechanical disruption (Thetford and Gilliam, 1991). Other challenges of hand weeding are the high cost of labor required (Gilliam et al., 1990). Sanitation and cultural practices along with the use of pre-emergence herbicides can help to reduce weeds in container-grown crops (Altland, 2003). Although pre-emergence herbicides are widely used during crop production, no pre-emergence herbicides are labeled for use in propagation and nursery growers remain reluctant using these products due to the safety issue on rooting (Cook and Neal, 2001; Langmaid, 1987). Also, pre-emergence herbicides cannot be used inside closed structures when crops are present. Since many crops are propagated inside closed structures, this limits the use of herbicides (Altland et al., 2003).

Mulches may be effective alternatives for controlling weeds in propagation. For container-grown crops, mulches are applied to the substrate surface to create a physical barrier which will inhibit weed seed germination and suppress weed growth (Ferguson et al., 2008). In an experiment by Ferguson et al. (2008), southern red cedar and southern magnolia wood chip mulches inhibited the germination of redroot pigweed (*Amaranthus retroflexus*) and large crabgrass (*Digitaria sanguinalis*) in nursery containers. Also, the wood chip mulches did not have an inhibitory effect on the growth of container-grown 'Carolina Beauty' crape myrtle (*Lagerstroemia indica* 'Carolina Beauty'). Nevertheless, some studies show that efficacy of mulches varies with the depth applied in a nursery container. Richardson et al. (2008) reported the reduction in oxalis (*Oxalis corniculata)* and bittercress numbers in large containers (#7) when pine bark mini nuggets were applied to a depth of either 3.8 or 7.62 cm (1.5 or 3 in.). In another study, flexuous bittercress (*Cardamine flexuosa*) and liverwort establishment and growth decreased with increasing rice hull depth (Altland et al., 2016). In an experiment by Altland and Krause (2014), containers with either a 1.3 or 2.5 cm (0.5 or 1.0 in.) depth of rice hulls provided nearly 100% weed control. In a separate study, pine bark mini nuggets at 2.5 cm (1 in.) depth reduced the germination of spotted spurge (*Chamaesyce maculata)* and eclipta (*Eclipta alba*) as compared to 1.3 cm (0.5 in.) mulch depth (Cochran et al., 2009).

Mulches may be an effective weed control method in crop propagation, but these products have not been thoroughly evaluated for cutting propagation. The objective of this research is to determine the effect of different mulches and application depth on rooting of stem cuttings and weed control in propagation.

MATERIALS AND METHODS

Two separate experiments (rooting stem cuttings and weed control) were conducted at the Tennessee State University, Otis L. Floyd Nursery Research Center in McMinnville, TN.

Rooting Stem Cuttings

Softwood cuttings (2-3 nodes) of hydrangea (*Hydrangea paniculata* 'Phantom') were collected from container-grown stock plants. Containers (6.6 cm diameter; SVD250, T.O. Plastics, Clearwater, MN) were filled with a 100% pine bark substrate amended with controlled-release fertilizer and micronutrient fertilizer. After saturation, mulches (coarse vermiculite, rice hulls, paper pellets, and pine pellets) were applied to containers at two depths [1.3 and 2.5 cm (0.5 and 1 in.)] and saturated. Cuttings were stuck (single cutting per container; 25 cuttings per treatment) on 19 May 2020. All cuttings received a 3-sec basal quick dip in rooting hormone (Dip'N Grow, Clackamas, OR) before sticking. Containers were completely randomized and placed under shade (50%) and intermittent mist (10 s every 8 min from 6:00 to 21:00). After 12 weeks, data were collected on rooting percentage, root dry weight, and digital image root analysis (total root length and root volume) using WinRhizo software (Reagent Instrument Canada Inc., Quebec City). Data were analyzed with linear models using the GLIMMIX procedure of SAS (Version 9.3; SAS Institute, Inc., Cary, NC, USA) and differences between treatment means were determined using the Shaffer-Simulated method ($P < 0.05$).

Weed Control

The weed control efficacy of four mulches applied at two depths (as described above) were evaluated on two weed species [large crabgrass (*Digitaria sanguinalis*) and bittercress (*Cardamine hirsuta*)]. Containers (6.6 cm diameter) were filled with substrate, placed under mist until saturation, then mulches were applied.

Twenty (bittercress) or 30 (crabgrass) seeds were sown per container (8 replications per treatment) on 28 February 2020. Containers were completely randomized (within species) and maintained in a shade house under intermittent mist (as described above). Weed seedling count was recorded at 2, 4, and 6 weeks after sowing (WAS). At 6 WAS, shoot fresh weight was collected. All data were analyzed as described above.

RESULTS

Rooting Stem Cuttings

Rooting percentage for hydrangea was 100% for all treatments except for rice hulls (90%) and paper pellets (95%) applied at 2.5 cm (1 in.) (Table 1). Root dry weight and total root length were similar to the non-treated control for all mulches. Root volume was lowest for paper pellets at 2.5 cm (1 in.) depth, but root volume for all the other treatments was similar to the non-treated control (Figs. 1 and 2).

Table 1. Rooting percentage, root dry weight, total root length and root volume of hydrangea cuttings treated with mulches at two depths.

Treatment	Mulch Depth (cm)	Rooting (%)	Root dry weight (g)	Total root length (cm)	Root volume (cm^3)
Non-treated control	NA	100 a[z]	0.16 a	771.3 ab	1.88 a
Vermiculite	1.3	100 a	0.17 a	847.6 a	2.06 a
Rice hulls		100 a	0.12 a	615.7 ab	1.41 ab
Pine pellets		100 a	0.16 a	762.5 ab	1.77 ab
Paper pellets		100 a	0.17 a	780.9 ab	1.66 ab
Vermiculite	2.5	100 a	0.18 a	872.6 a	2.11 a
Rice hulls		90 a	0.14 a	608.8 ab	1.43 ab
Pine pellets		100 a	0.17 a	793.2 ab	1.79 ab
Paper pellets		95 a	0.11 a	504.3 b	1.07 b

[z]Means followed by different letters within columns indicate significant difference at $P < 0.05$ using the Shaffer-Simulated method for multiple comparisons.

Figure 1. Hydrangea root and shoot growth after two months when treated with four mulches at 0.5- and 1-inch depth. Left to right: Control, vermiculite (0.5inch), rice hulls (0.5inch), pine pellets (0.5inch), paper pellets (0.5inch), vermiculite (1inch), rice hulls (1inch), pine pellets (1inch), and paper pellets (1inch).

Figure 2. Hydrangea cuttings under mist after treated with four mulches (Vermiculite, Rice hulls, Pine pellets and Paper pellets) at 0.5- and 1-in. depth.

Table 2. Seedling count of two weed species at 2, 4 and 6 weeks after sowing onto mulches at two depths.

		Crabgrass			Bittercress		
		Weed seedling count					
Treatment	Mulch depth (cm)	2 WAS	4WAS	6WAS	2WAS	4WAS	6WAS
Non-treated control	NA	11.3 b	16.8 ab	17.6 abc	16.5 a	19.3 a	18 a
Vermiculite		17 a	19.4 a	19.6 ab	18.3 a	18.8 a	17.8 a
Rice hulls		4.6 cd	15.5 abc	15.5 abc	9.9 b	17.9 a	16.8 a
Pine pellets	1.3	3.9 cd	14.9 abc	15.5 abc	2 cd	10.9 b	0.8 b
Paper pellets		7.9 bcd	14.8 abc	14.9 bc	3.6 c	17.1 a	19 a
Vermiculite		17.6 a	19.8 a	20.6 a	19.1 a	18.5 a	18.8 a
Rice Hulls	2.5	9.4 bc	15.9 ab	14.6 bc	11.4 b	17.4 a	17.1 a
Pine pellets		3.1 d	10.1 c	12.8 c	0.3 d	10.5 b	0 b
Paper pellets		4.1 cd	13.5 bc	13.9c	4.6 c	16.4 a	17.6 a

[z]Means followed by different letters within columns indicate significant difference at $P < 0.05$ using the Shaffer-Simulated method for multiple comparisons.

Weed Control

At 6 WAS, crabgrass seedling counts were similar for all treatments compared to the non-treated control (Table 2). Although crabgrass seed germination was not affected by the mulches, shoot fresh weight was suppressed at least 35% for all treatments except vermiculite at 2.5 cm (1 in.) depth, compared to the non-treated control (Table 3).

Bittercress seedling count (6 WAS) was lower for pine pellets at both mulch depths while all remaining treatments were similar to the non-treated control (Table 2). Shoot fresh weight of bittercress was 98% lower for paper pellets and pine pellets at both mulch depths compared to the non-treated control. For all other treatments, shoot fresh weight was similar to the non-treated control (Table 3).

Table 3. Shoot fresh weight of two weed species growing in containers treated with mulches at two depths.

Treatment	Mulch depth (cm)	Crabgrass	Bittercress
		Shoot Fresh Weight	(g)
Non-treated control	NA	4.39 a	0.97 ab
Vermiculite		3.12 ab	0.95 ab
Rice hulls	1.3	2.89 b	0.97 ab
Pine pellets		0.17 c	0.02 c
Paper pellets		0.02 c	0 c
Vermiculite		2.34 b	1.15 a
Rice Hulls	2.5	2.65 b	0.27 bc
Pine pellets		0.03 c	0 c
Paper pellets		0 c	0 c

[z]Means followed by different letters within columns indicate significant difference at $P < 0.05$ using the Shaffer-Simulated method for multiple comparisons.

DISCUSSION

The mulches used in this study did not affect root growth of hydrangea cuttings except paper pellets at 2.5 cm (1 in.) depth. In another study, 'Fashion' azalea (*Rhododendron indicum* x 'Fashion') had smaller growth indices when grown with paper pellets as a mulch (Smith et al., 1998). The chemical composition of the paper pellets is unknown but may have caused negative effects on root growth. The mulches used in our study were not very effective in lowering the germination of weed seeds, but several mulches were very effective in suppressing weed growth. Even though the seeds germinated, mulches prevented weed seedling from growing; hence, they did not compete with the crops. Pine pellets and paper pellets suppressed the growth of bittercress and crabgrass at both mulch depths, whereas vermiculite was only effective at suppressing crabgrass at 2.5 cm (1 in.) depth. In a study by Smith et al. (1998), recycled wastepaper pellets applied to a depth of 2.5 cm (1 in.) suppressed prostrate spurge (*Chamaesyce maculata)* germination.

Parboiled rice hulls have been shown to provide effective control of flexuous bittercress or creeping woodsorrel when applied 1.25 to 2.5 cm (0.5 and 1 in.) deep over the container substrate surface during crop production (Altland et al., 2016). In the present study, rice hulls did not provide adequate weed control, yet rice hulls are successfully used as mulch for controlling weeds in production. The hydrophobic nature of rice hulls creates dry environment, which inhibit the weed seed germination. But due to frequent irrigation in propagation rice hulls are moist enough creating a favorable condition for weed seed to germinate.

We concluded that among all the mulches used in the study, pine pellets and paper pellets were most effective in controlling weeds in propagation. Although paper pellets resulted in a slight reduction in root growth for hydrangea, this may be more acceptable compared to the risk of injury due to herbicide use and high cost of labor associated with hand weeding.

Literature Cited

Altland, J. (2005). Weed Control in Nursery Field Production. Extension Horticulture Faculty, North Willamette Research and Extension Center, Oregon State University. EM 8899-E.

Altland, J. and Krause, C. (2014). Parboiled rice hull mulch in containers reduces liverwort and flexuous bittercress growth. J. Environ. Hort. *32*:59-63.

Altland, J., Regan, R., and Newby, A. (2003). Liverwort control in propagation: Challenges and opportunities. Comb. Proc. Intl. Plant Prop. Soc. *53*:383-386.

Altland, J.E. (2003). Weed control in container crops: A guide to effective weed management through preventive measures. Oregon St. Univ. Coop. Ext. Publ. EM 8823.

Altland, J.E., Boldt, J.K., and Krause, C. (2016). Rice hull mulch affects germination of bittercress and creeping woodsorrel in container plant culture. Amer. J. Plant Sci. *7*:2359-2375.

Cochran, D.R., Gilliam, C.H., Eakes, D.J., Wehtje, G.R., Knight, P.R., and Olive, J. (2009). Mulch depth affects weed germination. J. Environ. Hort. *27*:85-90.

Cook, J.C. and Neal, J.C. (2001). Effects of herbicides and application timing on rooting of azalea and Japanese holly cuttings. Proc. SNA. Res. Conf. *46*:422-424.

Cranston, R. (1994). Weed control: An introductory manual. British Columbia Ministry of Agr. Fisheries and Food, Abbottsford.

Ferguson, J., Rathinasabapathi, B., and Warren, C. (2008). Southern redcedar and southern magnolia wood chip mulches for weed suppression in containerized woody ornamentals. HortTech.*18*:266-270.

Gilliam, C.H., Foster, W.J., Adrain, J.L., and Shumack, R.L. (1990). A survey of weed control costs and strategies in container production nurseries. J. Environ. Hort. *8*:133-135.

Hobbs, H.A., Johnson, R.R., Story, R.N., and Black, L.L. (1995). Weed hosts and thrips transmission of tomato spotted wilt virus in Louisiana. Tospoviruses and Thrips of Floral and Vegetable Crops. Acta Hortic. *431*:291-297.

Langmaid, M. (1987). Herbicides used in propagation at Wight Nurseries. Proc. Intern. Plant Prop. Soc. *37*:539-540.

Mathers, H.M. (2003). Novel methods of weed control in containers. HortTech. *13*:28-34.

Richardson, B.M., Gilliam, C.H., Wehtje, G.R., and Fain, G.B. (2008). Pine bark mini nuggets provide effective bittercress and oxalis control. J. Environ. Hort. *26*:144-148.

Smith, D.R., Gilliam, C.H., Edwards, J.J., Olive, J.W., Eakes, D.J., and Williams, J.D. (1998). Recycled wastepaper as a non-chemical alternative for weed control in container production. J. Environ. Hort. *16*: 69-75.

Thetford, M. and Gilliam, C.H. (1991). Herbicide use in propagation: Effects on rooting and root growth of stem cuttings. J. Environ. Hort. *9*:21-23.

Combined Proceedings IPPS 70:60-70. 2020.

Green Lacewing (*Chrysoperla rufilabris*) is a Voracious Predator on Crapemyrtle Bark Scale (*Acanthococcus lagerstroemiae*)

Bin Wu[1], Runshi Xie[1], Mengmeng Gu[2*], Hongmin Qin[3]

[1]Department of Horticultural Sciences, Texas A&M University, College Station, TX 77843, USA; [2]Department of Horticultural Sciences, Texas A&M AgriLife Extension Service, College Station, TX 77843, USA; [3]Department of Biology, Texas A&M University, College Station, TX 77843, USA.

mgu@tamu.edu ; hqin@bio.tamu.edu

Keywords: Biological control, green lacewings, predatory capacity, Y-tube assay

Summary

Crapemyrtle bark scale (CMBS, *Acanthococcus lagerstroemiae*), an invasive sap-sucking hemipteran, has spread across 14 states of the United States. The infestation of CMBS negatively impacted the flowering of some ornamental plants, and even the fruiting of some economically important crops. Using natural enemies, a non-chemical approach, would be beneficial for the integrated management of CMBS. Eggs of the green lacewings were observed on CMBS-infested crapemyrtle plants at Texas A&M University campus. Aiming to utilize green lacewings (*Chrysoperla rufilabris*) as a biocontrol agent of CMBS, predatory capacity of the green lacewings on CMBS was evaluated in laboratory conditions in this study. The results confirmed that the larval green lacewings could prey upon CMBS's nymphs, eggs, and adults. The average duration of the first egg consumption ($P<0.0001$) and the mean number of CMBS eggs consumed per larval green lacewing in 24 hours ($P<0.0001$) differed among different developmental stages. The 1st instar lacewing took 141.4 ± 4.8 sec. (mean ± SE) to consume the first CMBS egg and finished 11.8 ± 1.3 CMBS eggs in 24 hours. Whereas, the 3rd instar green lacewings devoured the first egg in 60.3 ± 3.0 seconds and consumed 176.4 ±

6.9 eggs per 24 hours. The Y-tube assay demonstrated that 78.1 ± 4.7% of larval *C. rufilabris* located CMBS under dark conditions. Thus, the evaluation of the predation capacity and Y-tube results confirmed that *C. rufilabris* could potentially be applied to control CMBS biologically.

INTRODUCTION

Crapemyrtle bark scale (CMBS, *Acanthococcus lagerstroemiae*) is a sap-sucking pest introduced from other countries (Gu et al., 2014; Merchant et al., 2014). Similar to aphids, crapemyrtle bark scale secretes honeydew when feeding on a plant. The infestation of CMBS seriously affects host plants' growth and development, even leading to branch die-back (Wang et al., 2019). This exotic pest was firstly reported in Texas in 2004 and has rapidly spread to 14 states, including Alabama, Arkansas, Georgia, Kansas, Louisiana, Mississippi, New Mexico, North Carolina, Oklahoma, South Carolina, Tennessee, Texas, Virginia (EDDMapS, 2020), and Washington (personal communication).

Crapemyrtle bark scale is a polyphagous insect and threatening many plants. In our previous study, different *Lagerstroemia* species and *Callicarpa* species (*L. indica* 'Dynamite', *L. fauriei* 'Kiowa', *L. limii*, *L. subcostata*, *L. speciosa*, *C. dichotoma* 'Issai', *C. americana* 'Bok Tower', *C. longissima* 'Alba', and *C. randaiensis*) were confirmed as CMBS hosts (Wu et al., 2019). The CMBS infestation was also observed on apple (*Malus domestica*), twelve pomegranate cultivars (*Punica granatum*), and other crop plants (Xie et al., 2020).

Natural enemies could be utilized to control CMBS biologically. In the field and laboratory conditions, cactus lady beetles (*Chilocorus cacti*) were confirmed as a predator on CMBS in Louisiana and Texas (Wang et al., 2016a; Wang et al., 2016b). We noticed that eggs of green lacewings (*Chrysoperla rufilabris*) were found feeding on CMBS-infested crapemyrtle plants at Texas A&M University campus (Fig. 1).

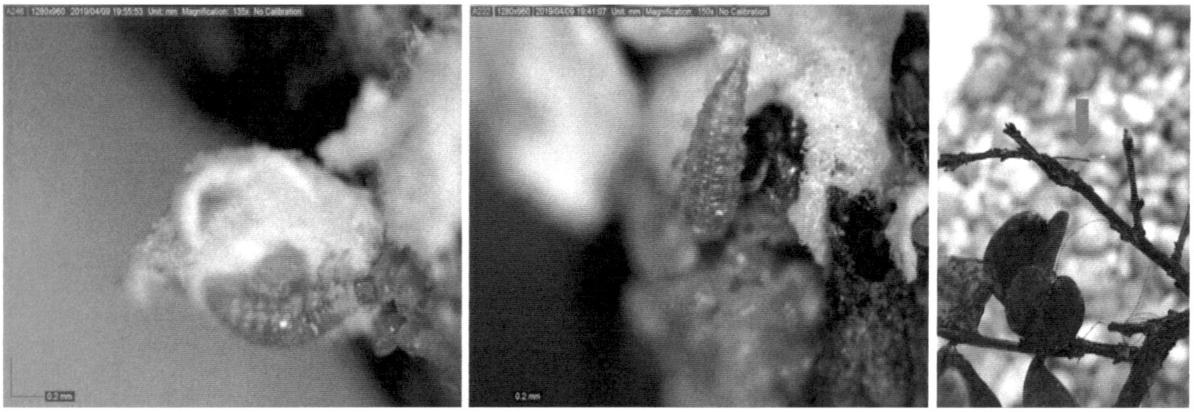

Figure 1. Several larval *Chrysoperla rufilabris* were observed feeding on CMBS female adults and lacewing eggs were found on CMBS-infested plants under natural environment.

Green lacewing is a highly fecund holometabolous insect, averaging 284 eggs per oviposition (Albuquerque et al., 1994). *Chrysoperla carnea* and *C. rufilabris* have been applied for pest biocontrol for many years in greenhouse and field crops (Tauber et al., 2000). The predation capacity of *C. rufilabris* on CMBS has not been reported yet.

Volatile communication is vital in moderating the interactions between insects, host plants, and other organisms (de Vos and Jander, 2010; Sudhida et al., 2010; Van Emden and Hagen, 1976; Zhu et al., 2005). Electroantennography (EAG) response was recorded from *C. carnea* in response to semiochemicals released from the prey (Zhu et al., 1999). Similarly, *Chrysoperla rufilbaris* searching CMBS may involve olfactory. Thus, foraging behavior and predatory capacity upon CMBS eggs in 24 hours was investigated under laboratory conditions to confirm *C. rufilabris* as an effective biocontrol agent of CMBS. A Y-tube assay was conducted to test if the lacewings could locate CMBS under dark.

MATERIALS AND METHODS

Insects. Larvae of *C. rufilabris* were purchased from ARBICO Organics™ (Oro Valley, AZ) and reared individually in Petri dishes (5.5 cm diameter) in a CONVIRON® -BDR 16 growth chamber (Controlled Environments Ltd., Winnipeg, Manitoba, Canada) at 25 ± 1 ℃ and 60 ± 5% relative humidity (R.H.) under a 12:12 (light: dark) photoperiod. An artificial diet containing 500mM sucrose, vitamins, minerals, and 150 mM amino acids (Prosser and Douglas, 1992) was used as food for the green lacewings. Larval *C. rufilabris* was held individually in a Petri dish (5.5 cm in diameter) and starved for 4hr before utilization in the tests. Additionally, the larvae were starved for 24 hours before their use in the June test.

Nymphs, adults, and eggs of CMBS were collected from naturally CMBS-infested crapemyrtle plants in College Station, TX.

Experiment 1. Can green lacewings prey on CMBS in lab conditions?

Crapemyrtle bark scale nymphs, adults, and eggs were distributed in a Petri dish (5.5 cm in diameter) to test whether *C. rufilabris* larvae would effectively prey on CMBS. Then, a larval *C. rufilabris* was placed in the same Petri dish. The preying behavior of the green lacewing was investigated using Stemi 2000 stereomicroscope (Carl Zeiss AG, Oberkochen, Germany) in laboratory conditions.

Experiment 2. How many CMBS eggs does a larval green lacewing consume in 24 hours?

The predatory evaluation experiment was conducted, respectively, in June and October 2019. In the June test, an individual 24h starved green lacewing was introduced into the center of a Petri dish (5.5 cm in diameter) containing approximately 300 fresh CMBS eggs. The duration when a larval *C. rufilabris* consumed the first CMBS egg entirely was recorded. After 24 hours, the number of CMBS eggs consumed by *C. rufilabris* was counted with the help of ImageJ (a Java-based image processing program developed at the National Institutes of Health and the Laboratory for Optical and Computational Instrumentation). An image of the eggs in the Petri dish before feeding was taken to easier compare and confirm the number of eggs consumed by the larval green lacewing in the 24 hours. The 24hr predation test was repeated six times, tracking the same individual *C. rufilabris* with a 24hr starvation as an interval. Thirteen effective repetitions were recorded and plotted using Excel.

In the October test, an individual 1st, 2nd, or 3rd instar larva of the starved green lacewing was inoculated into the center of a Petri dish (5.5 cm in diameter) containing approximately 300 fresh eggs of CMBS. The duration when lava consumed the first CMBS egg completely was recorded. After 24 hours, the number of CMBS eggs consumed by *C. rufilabris* was counted with the help of ImageJ. A set of twenty dishes was replicated three times for each instar. For the data analysis, the duration of the first egg consumption and the prey consumption of the green lacewing were analyzed using repeated-measures ANOVA.

Stages, replicates, the stages*replicates interaction were the variables, assigned subjects nested with stages as a random effect. The means of the duration and consumption per 24hr were, respectively, separated using Tukey's HSD test at α =0.05 (JMP Pro 15, SAS Institute, Cary, NC).

Experiment 3. Are green lacewings able to forage CMBS in the dark?

A Y-tube assay was set up in this study (Fig. 2). Three glass vials were joined by a Bel-Art Y-tubing connector (SP Scienceware, Wayne, NJ). The loading vial contained a starved second or third instar green lacewing, the baited vial contained over ten alive gravid females and some crawlers, and the control vial was vacant. The Y-tube setting was boxed to avoid visual stimuli, and the box was put into the CONVIRON® -BDR 16 growth chamber at 25±1°C and 60±5% R.H. under a 24-hour dark photoperiod.

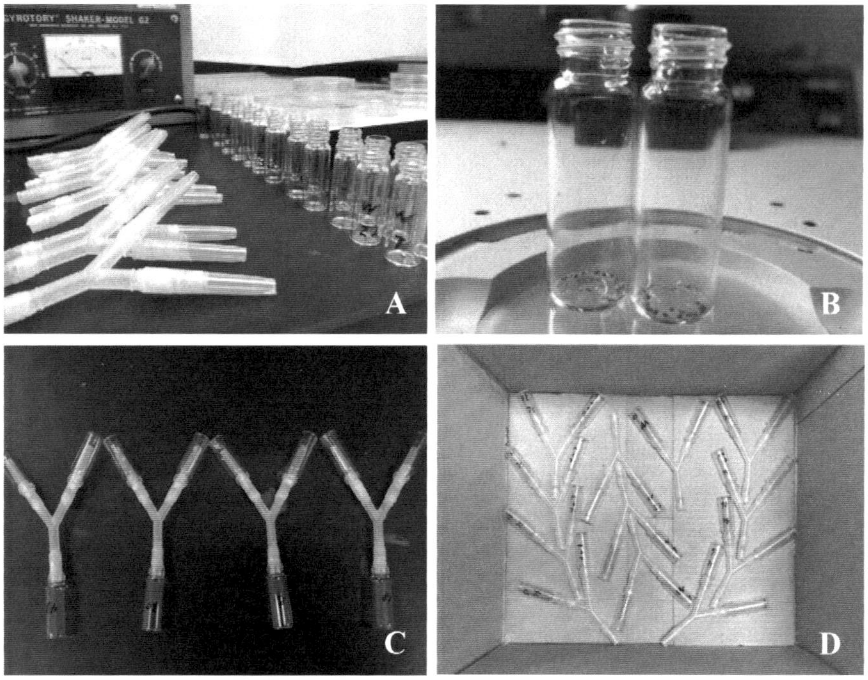

Figure 2. Y-tube test assemble. A: Y-tube tubes and vials; B: Vials containing CMBS; C: Each Y-tube set up was assembled by a Y tubing connector (6.0 mm tubing I.D.) and contained a loading vial, a baited vial and a control vial; D: Y-tube tests were conducted under dark.

Twelve Y-tube settings were performed simultaneously. After 24 hours, the number of predators that entered the baited vials (B) and the control vials (C) was counted. The positive response ratio (%) was calculated as B/(B+C) *100, and the negative response ratio (%) was calculated as C/(B+C) *100. The test was repeated ten times using fresh CMBS and larval green lacewings. The data were analyzed using one-way ANOVA (JMP Pro 15, SAS Institute, Cary, NC), and the response ratios were separated using Tukey-Kramer HSD ($\alpha = 0.05$).

RESULTS AND DISCUSSION

Prey trial of larval *C. rufilbaris* on CMBS in laboratory conditions. The larval green lacewing was observed to devour gravid females and eggs of CMBS in the Petri dish (Fig. 3 and 4). Moreover, it was able to grab and devour tiny crawling nymphs under the lab conditions (Fig. 4). Together with observation in landscapes, green lacewings are probably natural biological control agents of CMBS.

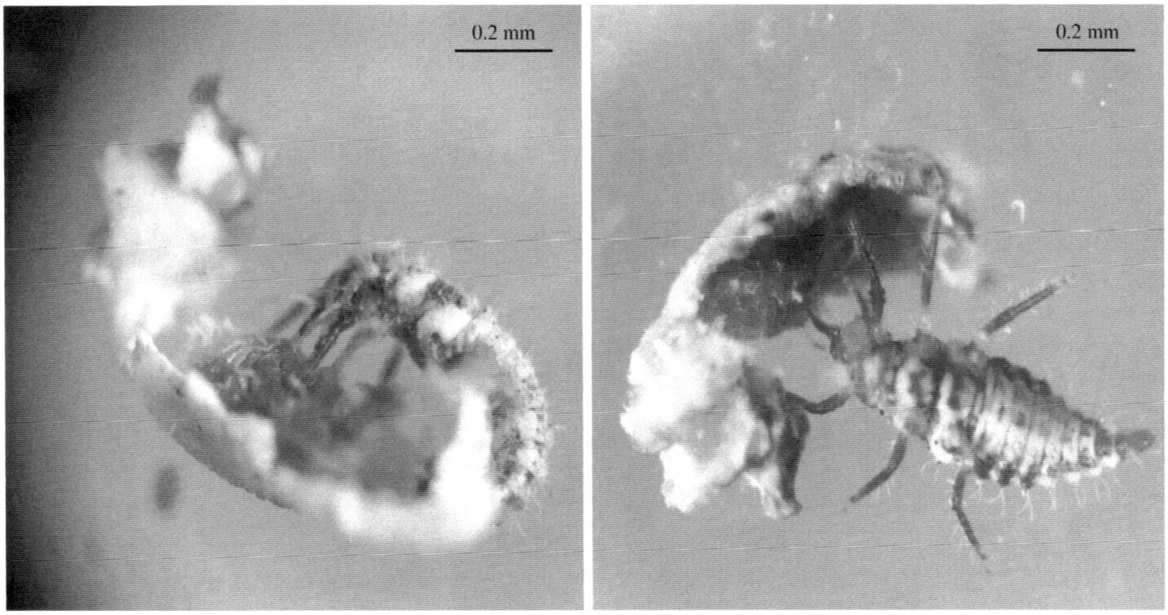

Figure 3. Nymphs of *Chrysoperla rufilabris* were preying on female adults of CMBS under laboratory condition**.** A: The larval green lacewing easily grasped and voraciously attacked a female adult of CMBS by seizing it with its large, sucking jaws after placing them together in the same petri dish; B: The larva of *C. rufilabris* esuriently consumed the body fluids of the CMBS female leading to the CMBS shrinking and death.

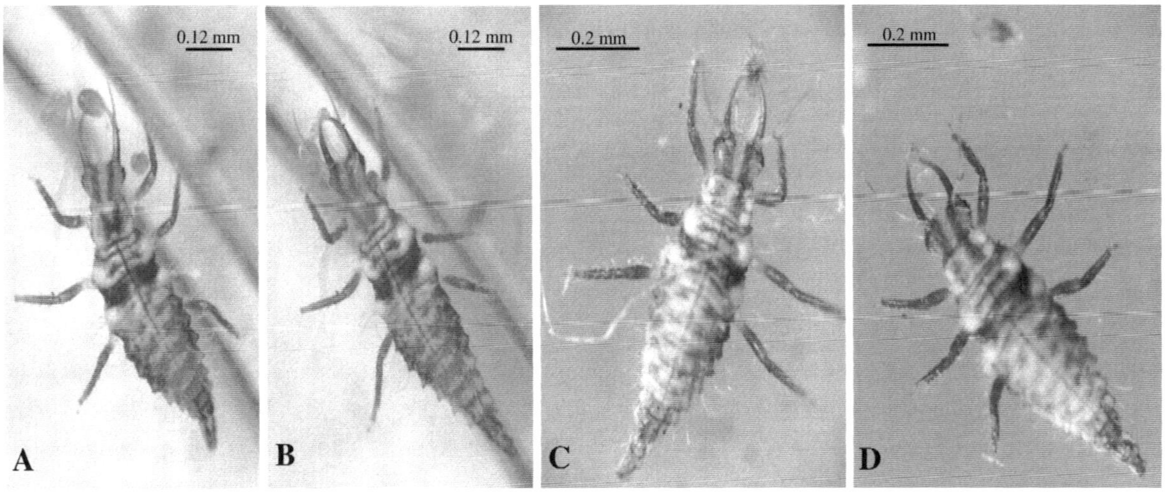

Figure 4. Larval green lacewings were able to prey on eggs and crawlers of CMBS under laboratory conditions**.** A: The larval *C. rufilabris* grasped a CMBS egg; B: The larva consumed the egg in around one minute after grabbed it; C: A larva of *C. rufilabris* seized a CMBS crawler; D: The larva rapaciously pierced the crawler.

Evaluation of the predatory capacity upon CMBS eggs. In the June test (Fig. 5), the duration of the first egg consumption ranged from 53.2 ± 2.5 seconds to 73.2 ± 2.7 seconds. The average number of CMBS eggs that the larval green lacewing consumed ranged from 154.1 ± 2.7 to 195.5 ± 2.5. In the October test, the analysis result of fixed effect tests showed that only the developmental stages of *C. rufilabris* impacted the duration of the first egg consumption ($P<0.0001$) and the number of CMBS eggs eaten in 24 hours ($P<0.0001$).

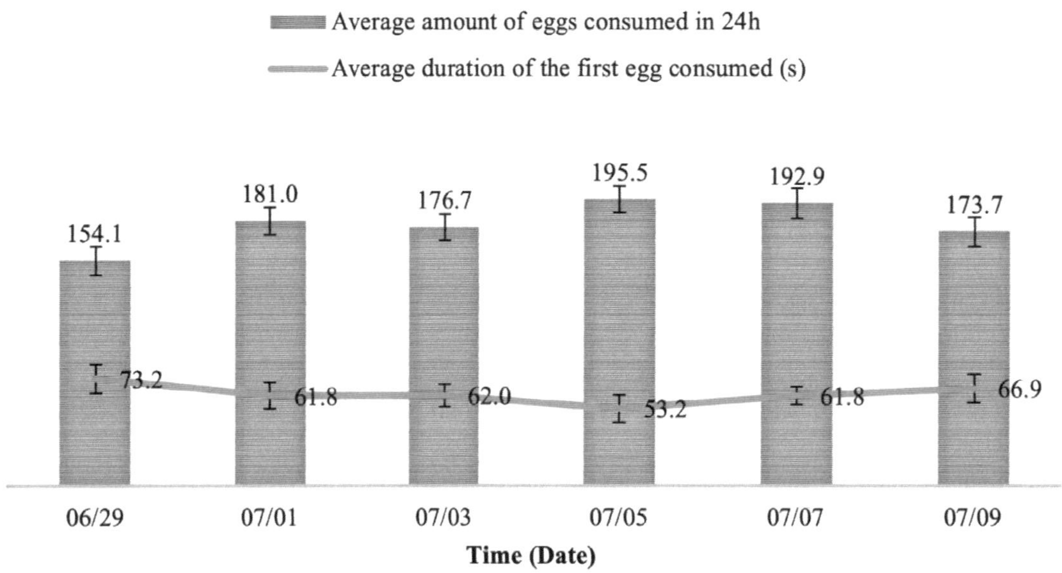

Figure 5. 12-day tracking results on the mean amount of CMBS eggs consumed by a larval *Chrysoperla rufilabris* in every 24 hours (blue columns) and the average duration of the first CMBS egg consumed (orange line).

Subsequently, the three-replicate data for each development stage, 60 dishes as a set, were calculated and shown in Table 1. The average duration of the first egg consumption dropped from 141.4 ± 4.8 seconds (mean ± SE) in the 1st instar to 60.3 ± 3.0 seconds in the 3rd instar. The average number of CMBS eggs consumed by a larval green lacewing in 24 hours increased dramatically from 11.8 ± 1.3 in the 1st instar to 176.4 ± 6.9 in the 3rd instar.

The results in Experiment 1 demonstrated that larval green lacewing voraciously preyed on crawling nymphs, adults, and eggs of CMBS. However, similar to lady beetle species (Bilde and Toft, 1997; Bilde and Toft, 1999; Finlayson et al., 2010; Golizadeh and Jafari-Behi, 2012; Omkar and Sahu, 2009), green lacewings showed significantly feeding preference among aphid species (Chen and Liu, 2001; Liu and Chen, 2001).

Table 1. The effect of developmental stages on prey consumption of *Chrysoperla rufilabris* upon CMBS eggs.

Developmental stage	Number of larval green lacewings tested	Replicate	Duration of the first egg consumption (seconds ± S.E.)	Mean number of CMBS consumed in 24 hours (± S.E.)
1st instar	20	3	141.4 ± 4.8a	11.8 ± 1.3c
2nd instar	20	3	77.5 ± 4.7b	151.5 ± 6.6b
3rd instar	20	3	60.3 ± 3.0c	176.4 ± 6.9a

* Means, in the same column, followed by different letters are significantly different (P<0.05) as determined by Tukey's HSD test.

The 3rd instar of the lacewings can consume at most 277 CMBS eggs in Experiment 3, compared with approximately 400 scale eggs eaten by the 4th instar of *C. cacti* in 24 hours under the lab conditions (Wang et al., 2016a). Thus, to select the optimal predator in effectively controlling CMBS, it would be necessary and interesting to compare the predation capacity on CMBS between some lady beetle species and the green lacewing species.

Olfactory response using a Y-tube assay. The analysis results showed that the positive response ratio was significantly higher than the negative ratio (P<0.0001), which indicated that the lacewings were attracted to CMBS in the Y-tube settings. In detail, 78.14 ± 4.74% of the larval green lacewings were able to locate CMBS in the Y-tube assay under dark (Fig. 6).

Electroantennography responses to some sex pheromone components or alarm pheromones of the aphids (Homoptera: Aphididae) have been obtained using green lacewings, *Chrysopa cognata* (Boo et al., 1998; Cho et al., 2014) and *Chrysopa pallens* (Li et al., 2017), and other natural enemies like *Coccinella septempunctata* (Al Abassi et al., 2000) and *Adalia bipunctata* (Francis et al., 2004). Some pheromone compounds were investigated and proved influential in the field trapping experiment (Boo et al., 2003; Zhang et al., 2006). Similarly, spraying the extraction of CMBS volatile compounds could be technologically helpful to enhance the efficiency of CMBS biocontrol before the infestation happens heavily on a plant. Thus, it would be beneficial to focus on confirming and extracting the specific compounds in the future study.

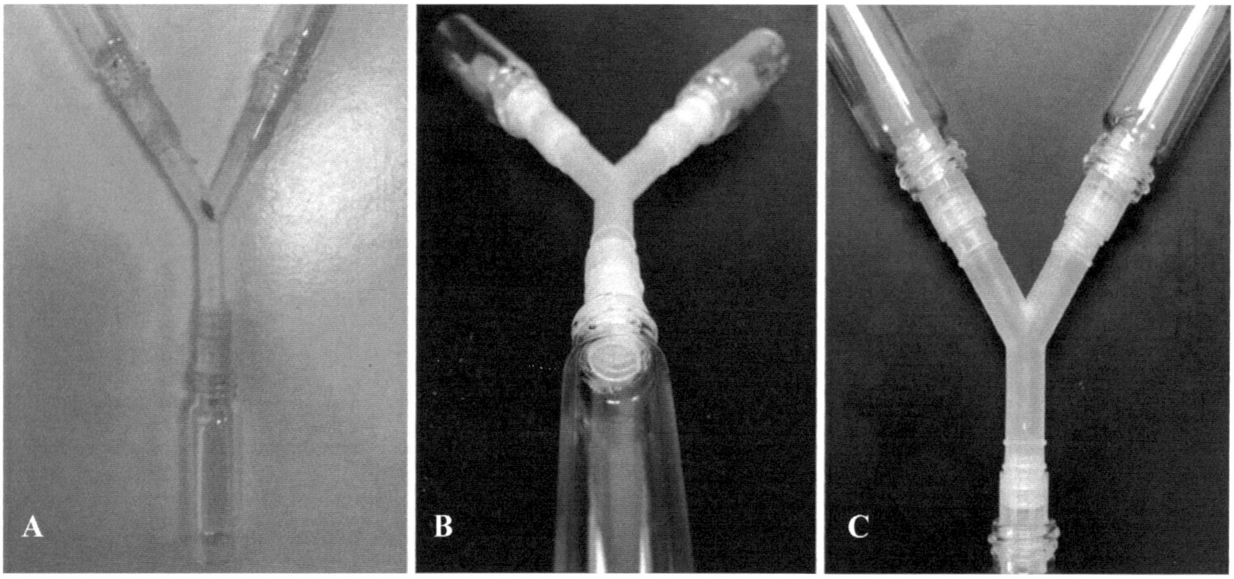

Figure 6. Different responses in the Y-tube test under dark. After 24 hours, some larval green lacewings went to the control vials (A) or stayed in the loading vials (B) Still, 78.14 ± 4.74% of the larval *C. rufilabris* were able to locate CMBS in the baited vials.

Acknowledgments

This work is supported by TAMU T_3 246495-2019, Crop Protection and Pest Management project 'Integrated pest management strategies for crape myrtle bark scale, a new exotic pest' (No. 2014-70006-22632) and Specialty Crop Research Initiative project 'Systematic Strategies to Manage Crapemyrtle Bark Scale, An Emerging Exotic Pest' (grant no. 2017-51181-26831) from the USDA National Institute of Food and Agriculture. Any opinions, findings, conclusions, or recommendations expressed in this publication are those of the authors and do not necessarily reflect the view of the U.S. Department of Agriculture.

Literature Cited

Al Abassi, S., Birkett, M., Pettersson, J., Pickett, J., Wadhams A.L., and Woodcock, C. (2000). Response of the seven-spot ladybird to an aphid alarm pheromone and an alarm pheromone inhibitor is mediated by paired olfactory cells. J. Chem. Ecol. *26*:1765-1771.

Albuquerque, G.S., Tauber, C.A., and Tauber, M.J. (1994). *Chrysoperla externa* (Neuroptera: Chrysopidae): life history and potential for biological control in Central and South America. Biolog. Control 4:8-13.

Bilde, T., and Toft, S. (1997). Consumption by carabid beetles of three cereal aphid species relative to other prey types. Entomophaga *42*:21-32.

Bilde, T., and Toft, S. (1999). Prey consumption and fecundity of the carabid beetle *Calathus melanocephalus* on diets of three cereal aphids: High consumption. Pedobiologia *43*:429.

Boo, K., Chung, I., Han ,K., Pickett, J., and Wadhams, L. (1998). Response of the lacewing *Chrysopa cognata* to pheromones of its aphid prey. J. Chem. Ecol. *24*:631-643.

Boo, K.S., Kang, S., Park, J., Pickett,J., and Wadhams, L. (2003). Field trapping of *Chrysopa cognata* (Neuroptera: Chrysopidae) with aphid sex pheromone components in Korea. J. Asia-Pacific Entomol. *6*:29-36.

Chen, T., and Liu, T. (2001). Relative consumption of three aphid species by the lacewing, *Chrysoperla rufilabris*, and effects on its development and survival. BioControl *46*:481-491.

Cho, J., Lee, M., Park, C., Kim, J., Hooper, T., Woodcock, C., and Pickett, J. (2014). Behavioral response of the lacewing *Chrysopa cognata* to both *Aphis gossypii*-induced plant volatiles and *Chrysopa cognata*-derived volatiles. Korean J. Appl. Entomol. *53*:7-13.

de Vos M., and Jander, G. (2010). Volatile communication in plant–aphid interactions. Cur. Opin. Plant Biol. *13*:366-371.

EDDMapS. (2020). Early Detection & Distribution Mapping System Available online at http://www.eddmaps.org/; last accessed August 23, 2020.

Finlayson, C., Alyokhin, A., Gross, S., and Porter, E. (2010). Differential consumption of four aphid species by four lady beetle species. J. Insect Sci. *10*.

Francis, F., Lognay, G., and Haubruge, E. (2004). Olfactory responses to aphid and host plant volatile releases:(E)-β-farnesene an effective kairomone for the predator Adalia bipunctata. J. Chem. Ecol. *30*:741-755.

Golizadeh, A., and Jafari-Behi, V. (2012). Biological traits and life table parameters of variegated lady beetle, *Hippodamia variegata* (Coleoptera: Coccinellidae) on three aphid species. Appl. Entomol. Zool. *47*:199-205.

Gu, M., Merchant, M., Robbins, J., and Hopkins, J. (2014). Crape myrtle bark scale: A new exotic pest. Texas A&M AgriLife Ext. Service. EHT 49.

Li, Z.Q., Zhang, S., Cai, X.M., Luo, J.Y., Dong, S.L., Cui, J.J., and Chen, Z.M. (2017). Three odorant binding proteins may regulate the behavioural response of *Chrysopa pallens* to plant volatiles and the aphid alarm pheromone (E)-β-farnesene. Insect Molecul. Biol. *26*:255-265.

Liu, T., and Chen, T. (2001). Effects of three aphid species (Homoptera: Aphididae) on development, survival and predation of *Chrysoperla carnea* (Neuroptera: Chrysopidae). Appl. Entomol. Zoolog. *36*:361-366.

Merchant, M.E., Gu, M., Robbins, J., Vafaie, E., Barr, N., Tripodi, A.D., and Szalanski, A.L. (2014). Discovery and spread of *Eriococcus lagerstroemiae* Kuwana (Hemiptera: Eriococcidae), a new invasive pest of crape myrtle, *Lagerstroemia* spp.

Omkar, K.G., and Sahu, J. (2009). Performance of a predatory ladybird beetle, *Anegleis cardoni* (Coleoptera: Coccinellidae) on three aphid species. Europ. J. Entomol. *106*:565-572.

Prosser, W., and Douglas, A. (1992). A test of the hypotheses that nitrogen is upgraded and recycled in an aphid (*Acyrthosiphon pisum*) symbiosis. J. Insect Physiol. *38*:93-99.

Sudhida, G., Singh, A.K., and Gautam, R.D. (2010). Olfactory responses of green lacewings, *Chrysoperla* sp.(carnea group) and *Mallada desjardinsi* on mealybug, *Phenacoccus solenopsis* (Homoptera: Pseudococcidae) fed on cotton. Acta Entomol. Sinica *53*:497-507.

Tauber, M.J., Tauber, C.A., Daane, K.M., and Hagen, K.S. (2000). Commercialization of predators: recent lessons from green lacewings (Neuroptera: Chrysopidae: Chrosoperla). Amer. Entomol. *46*:26-38.

Van Emden, H., and Hagen, K. (1976). Olfactory reactions of the green lacewing, *Chrysopa carnea*, to tryptophan and certain breakdown products. Environ. Entomol. *5*:469-473.

Wang, Z., Chen, Y., and Diaz, R. (2016a). The cactus lady beetle: a voracious predator of scale insects.

Wang, Z., Chen, Y., and Diaz, R. (2019). Temperature-dependent development and host range of crapemyrtle bark scale, *Acanthococcus lagerstroemiae* (Kuwana)(Hemiptera: Eriococcidae). Florida Entomol. *102*:181-186.

Wang, Z., Chen, Y., Gu, M., Vafaie, E., Merchant, M., and Diaz, R. (2016b). Crapemyrtle bark scale: a new threat for crapemyrtles, a popular landscape plant in the US. Insects *7*:78.

Wu, B., Xie, R., Knox, G., Dinkins, M., Qin, H., and Gu, M. (2019). An Exotic Insect Pest, Crapemyrtle Bark Scale (*Acanthococcus lagerstroemiae*) (Hemiptera: Eriococcidae): Host Range and Acceptance Among 19 Plant Species *Intl. Plant Prop. Soc.- South. Reg. N. Amer.* (2019).

Xie, R., Wu, B., Dou, H., Liu, C., Knox, G.W., Qin, H., Gu, M. (2020). Feeding Preference of Crapemyrtle Bark Scale (*Acanthococcus lagerstroemiae*) on Different Species. Insects *11*:399.

Zhang, Q.-H., Sheng, M., Chen, G., Aldrich, J.R., and Chauhan, K.R. (2006). Iridodial: a powerful attractant for the green lacewing, *Chrysopa septempunctata* (Neuroptera: Chrysopidae). Naturwissenschaften *93*:461.

Zhu, J., Cossé, A.A., Obrycki, J.J., Boo, K.S., Baker, T.C. (1999). Olfactory reactions of the twelve-spotted lady beetle, *Coleomegilla maculata*, and the green lacewing, *Chrysoperla carnea* to semiochemicals released from their prey and host plant: electroantennogram and behavioral responses. J. Chem. Ecol. *25*:1163-1177.

Zhu, J., Obrycki, J., Ochieng, S.A., Baker, T.C., Pickett, J., and Smiley, D. (2005). Attraction of two lacewing species to volatiles produced by host plants and aphid prey. Naturwissenschaften *92*:277-281.

Combined Proceedings IPPS 70:71-80. 2020.

Biological Parameters of Crapemyrtle Bark Scale (*Acanthococcus lagerstromiae*) on *Lagerstroemia* 'Tuscarora' and Seedlings of 'Natchez' and 'Fantasy'

Runshi Xie[1], Bin Wu[1], Gary W. Knox[2], Mengmeng Gu[3], Hongmin Qin[4]

[1]Department of Horticultural Sciences, Texas A&M University, College Station TX 77843, USA; [2]Department of Environmental Horticulture, University of Florida/IFAS North Florida Research and Education Center, Quincy, FL 32351; [3]Department of Horticultural Sciences, Texas A&M AgriLife Extension Service, College Station, TX 77843, USA; [4]Department of Biology, Texas A&M University, College Station 77840, USA

fushe001@tamu.edu

Keywords: CMBS, crapemyrtle, ecology, life table, scale insect

Summary

Crapemyrtle bark scale [(CMBS); *Acanthococcus lagerstroemiae*], an exotic pest insect in the United States, causes damage to popular crapemyrtle landscape plants - as well as other economically important or native plant species, such as pomegranate, apple, and American beauty-berry. Age-stage, two-sex table study analysis was conducted to evaluate the biological parameters of CMBS on different species and cultivars of *Lagerstroemia* under laboratory conditions at 25° C and 250 μmol m^{-2} s^{-1} light with a photoperiod of 12:12 (light: dark). Crapemyrtle bark scale development was found to be greatly influenced by plant host. This study aimed to provide important biological and ecological data of CMBS. A comprehensive life table study was conducted for the first time - in order to gain a thorough understanding of CMBS development, survival, and fecundity on different plant species and cultivars of *Lagerstroemia*.

INTRODUCTION

Lagerstroemia spp. (Myrtatles: Lythraceae), commonly known as crapemyrtle, is a genus consisting at least 80 known species of trees or shrubs (Cabrera, 2002). Despite many cultivation purposes including timber production (Knox, 2000) and medicinal usage (Al-Snafi, 2019), crapemytles are mostly valued as ornamental plants with their versatile landscape use. Crapemytles are originally native species in tropical and subtropical regions of southeastern Asia (Pooler, 2007). Since its introduction to the western world in the 1600s, worldwide breeding and cultivation has resulted in more than 200 cultivars in the United States, Europe, Australia, and Asia (Pooler, 2007). The majority of naturalized or commercially available crapemyrtle cultivars in the U.S. are the selections made from *L. indica* seedlings, or progenies from hybridizations between *L. indica* and *L fauriei* (Wang et al., 2011).

Crapemyrtles are praised for various growth patterns and plant architectures, flower color and duration, attractive bark features, as well as disease and pest tolerance (Knox, 2000). The plant height of crapemyrtles ranges from 1 to 6.1m (3 to 20 ft) (Wade and Williams-Woodward, 2009), which makes it highly suitable for a variety of urban settings. The flowers are often the defining characteristics of crapemyrtle cultivars, which are determined by the flower colors and various types of inflorescences (Pooler, 2007). Many years of breeding effort and interspecific hybridization have produced cultivars with a wide range of flower colors, including red, purple, white, and other combined variants such as pink and lavender (Wang et al., 2010). However, the aesthetic value of crapemyrtles could be greatly undermined by serious infestation of crapemyrtle bark scale [(CMBS); *Acantho-coccus lagerstroemiae*] (Borchsenius, 1960).

Acanthococcus lagerstroemiae, native to East Asia, was originally categorized under the genus *Eriococcus*, and it was first described from specimens of adult females collected by Kuwana in Japan in 1907 (Kuwana, 1907). The binomial name *Acanthococcus lagerstroemiae* was first used by Borchsenius in 1960, where he also provided description and keys based on the morphology of adult female sample collected in China (Borchsenius, 1960b). Since its introduction to the United States, the infestation of this scale insect has been distributed widely and confirmed in at least 13 states (EDDMapS, 2019; Wang et al., 2016).

Heavy infestation caused by CMBS could hinder plant performance, which is associated with slow and weakened plant growth due to the active phloem-feeding by the insect, and/or reduced photosynthesis resulting from black sooty mold development on accumulated CMBS honeydew secretion on leaves. The damage caused by CMBS is not restricted to crapemyrtle, as economic plants such as apple, soybean, pomegranates, and native plant species such as American beautyberry and *Heimia* are also suitable hosts of CMBS (Wu; Xie et al., 2020).

In order to better manage and control CMBS, important ecological information regarding the development and life cycle of this pest is needed. However, previous studies are unclear about basic ecological data such as developmental stages of CMBS, as incongruent reports on CMBS life history were found. For example, Zhang reported only two instars before the female become sexually mature (Zhang and Shi, 1986), while other studies claimed the existence of the 3rd instar in female development (Jiang and Xu, 1998; Jiao and Zhang, 2011).

The study of the biology and life table is a system of record keeping, as well as analytical and mathematical approaches, to collect and interpret the data of an insect population (Harcourt, 1969). The construction of life tables enhances knowledge foundation and provides insight of the population dynamic of a certain arthropod of interest. However, such information for CMBS is currently lacking.

This study on the biology of CMBS focused on acquiring critical biological information in terms of a fertility life table, in order to expand the current knowledge related to the interaction between CMBS and its host plants in *Lagerstroemia*. This is needed for development of an effective integrated pest management (IPM) program in controlling CMBS.

MATERIALS AND METHODS

Insect source and handling

Branches/twigs infested with crapemyrtle bark scale were collected from crapemyrtle trees on campus (Texas A&M University, College Station, TX), and stored in zip-lock bags under constant temperature (25°C). White coverings of the female scales were carefully lifted using a fine pin/needle. All existing eggs inside the ovisacs were removed, and the gravid females were transferred, using a fine brush, onto a moist filter paper placed in a petri dish. All newly laid eggs were collected the next day (after 24 hours) and kept under 25°C for incubation until hatched.

Plant material and insect rearing chamber

Lagerstroemia species and cultivars, including *L.* 'Tuscarora' or the seedlings of *L.* 'Natchez' and *L.* 'Fantasy', were used as host/food sources for the CMBS rearing experiment.

Rearing chambers were constructed with small petri dishes (Falcon® Disposable Petri Dishes, 60 mm x 15 mm) and clear plastic food wrap. Around half of petri dish was wrapped by clear plastic food wrap to create space for the medium. Water agar (1%) was poured into the bottom of the petri dish, using an electronic pipette, to fill around one third portion of the petri dish. Stem cuttings with bud nodes were collected from different *Lagerstroemia* species and cultivars and stuck in agar medium to stay turgid. The rearing chambers were placed in Conviron growth chambers set at 25° C and 250 μmol m^{-2} s^{-1} light with a photoperiod of 12:12 (light: dark).

CMBS rearing experiment

Egg incubation times were recorded, which was considered from the day when the eggs were first laid by gravid females to the day 1st instars were hatched. One or two newly hatched crawlers/nymphs, per rearing chamber, were transferred onto stem cuttings using a fine brush. Daily observations were made to record the settling status for nymphs. Rearing chambers with nymphs that failed to settle on the plants were discarded when the mortality or escaping were confirmed.

Insect rearing experiments on different *Lagerstroemia* were conducted from April to December 2019. Daily observations were made as nymphs start feeding, and the duration of each developmental stage (including nymphal stages, pupa, and adult stages) were recorded. When a male reached adult stage, it was transferred to pair with a female for mating in order to complete the life cycle of the female. Fecundity data (the number of eggs that an adult female produces), and longevity (the number of days a female lives) were recorded as the gravid females complete their life cycle.

Data analysis

The developmental stages of both male and female CMBS were determined by the number of times the nymphs molt, which were obtained by keeping track of the exuviates. According to the age-stage, two-sex life table theory (Chi and Liu, 1985; Chi, 1988), the fecundity data (number of eggs each adult female produced) and longevity data (the number of days each CMBS nymph lives) can be obtained to calculate population (life table) parameters of CMBS.

To obtain population (life table) parameters, the life history data of CMBS was analyzed using TWOSEX-MS Chart, a computer program for the age-stage, and two-sex life table analysis (Chi, 2020), according to the method described in Chi and Su (2006). The raw data were used to calculate the age-stage specific survival rate (s_{xj}, where x = age in days and j = stage; the first stage is egg, the second stage is 1st instar, the third stage is 2nd instar, the fourth stage is male pupa 1, the fifth stage is male pupa 2, the sixth stage is male pupa 3, the seventh and eighth stages are female and male), age-specific survival rate (Sxj), and population (life table) parameters, including mean generation time (T), net reproduction rate (R_o), the intrinsic rate of increase (r), and the finite rate of natural increase (λ), to construct the age-stage, two-sex life table. The biological parameters and population parameters of CMBS reared on different *Lagerstroemia* hosts were compared using Student's t test and All Pair, Tukey HSD with JMP software (JMP Pro15, Statistical Analysis System, Cary, NC, USA).

RESULTS AND DISCUSSION

The development of a male consists of egg, two nymphal stages (1st and 2nd instar), three different stages of pupa and the winged adult stage. The development of a female undergoes four major stages: egg, two nymphal stages (1st and 2nd instar) and adult stage. The egg incubation time for both males and females are a little over 12 days at 25 °C. Crapemyrtle bark scale showed variable development times based on different crapemyrtle species and cultivars. For example, the average development duration of the male 1st instar was 15.0 ± 1 days (± SE) and 18.4 ± 0.7 days (± standard error) on 'Natchez' seedling and 'Fantasy' seedling, respectively (Table 1). No male had developed into adult on 'Tuscarora' during this study (Fig. 1). For the females, the average development duration was 32.3 ± 2.7 days (± SE), 12.8 ± 0.6 days (± SE), and 19.5 ± 1.4 days (± SE) on 'Tuscarora', 'Natchez' seedling, and 'Fantasy' seedling, respectively (Table 2).

Table 1. Means ± standard errors, and sample size of development duration of males of *Acanthococcus lagerstroemiae* at laboratory conditions of 25° C and 250 μmol m^{-2} s^{-1} light with a photoperiod of 12:12 (light:dark).

	Development duration (days), males						
	Natchez seedling		Fantasy seedling				
Stage	Mean ± SEM	N	Mean ± SEM	N	*t*	df	*P*
1st instar	15.0 ± 1	3	18.4 ± 0.7	5	2.93	6	0.027
2nd instar	21.1 ± 7.4	3	64.2 ± 13.2	5	2.35	6	0.057
Pupa1	4.7 ± 0.3	2	5.4 ± 0.4	5	1.25	6	0.258
Pupa2	3.5 ± 0.5	2	3.6 ± 0.4	5	0.14	5	0.895
Pupa3	4.5 ± 1.5	2	7.8 ± 1.7	5	1.15	5	0.304

There are significant differences at 0.05 confidence level in 1st instar duration between *A. lagerstroemiae* reared on 'Natchez' seedlings and 'Fantasy' seedlings according to Student's *t*-test.

Table 2. Means ± standard errors, and sample size of development duration of females of *Acanthococcus lagerstroemiae* at laboratory conditions of 25° C and 250 μmol m^{-2} s^{-1} light with a photoperiod of 12:12 (light: dark).

	Development duration (days), females								
	Tuscarora		Natchez seedling		Fantasy seedling				
Stage	Mean ± SEM	N	Mean ± SEM	N	Mean ± SEM	N	*F*	df	*P*
1st instar	32.3 ± 2.7 a[Z]	3	12.8 ± 0.6 c	4	19.5 ± 1.4 b	6	21.73	2,9	<0.01
2nd instar	14.5 ± 2.5 a	2	39.8 ± 18.7 a	4	62.8 ± 4.6 a	6	3.554	2,9	0.07

[Z] Means within each row followed by the same letter are not significantly different according to All Pairs, Tukey Honestly Significant Difference at 0.05 confidence level.

According to the life table analysis of CMBS on different *Lagerstroemia*, the insect population developed fastest on 'Natchez' seedlings, with shortest mean generation time (*T*) of 41.88 ± 21.07 days (± SE), compared to the longest *T* (103.77 ± 14.42 days ± SE) found on 'Fantasy' seedlings (Table 3), which suggests that the development rate of CMBS infestation could be nearly tripled on different hosts. The lowest fecundity of CMBS was also observed on 'Fantasy' seedlings (22.07±10.22 offspring per individual), which was relatively low as compared to the highest mean fecundity found on 'Natchez' seedlings (69.97±43.26 offspring per individual) (Table 3).

The development and survival of CMBS was also greatly influenced when reared on different species or cultivars of crapemyrtles (Figs 1, 2 and 3), which suggests that the population dynamics of CMBS could vary drastically based upon the host selection.

Table 3. Intrinsic rate of increase (r) ± standard errors, finite rate of increase (λ) ± standard errors, net reproduction rate (R_o) ± standard errors, mean generation time (T) ± standard errors, and gross reproductive rate (GRR) ± standard errors of *Acanthococcus lagerstroemiae* at laboratory conditions of 25° C and 250 µmol m^{-2} s^{-1} light with a photoperiod of 12:12 (light: dark).

Plants	N	r (days^{-1})	λ	Ro	T (days)	GRR (offspring/individual)
Tuscarora	10	0.04 ± 0.009 b[Z]	1.04 ± 0.009 b	8.95 ± 4.38 b	50.33 ± 0.02 b	39.81 ± 15.38 b
Natchez seedling	15	0.09 ± 0.04 a	1.09 ± 0.04 a	21.42 ± 11.05 a	41.88 ± 21.07 c	69.97 ± 43.26 a
Fantasy seedling	35	0.01 ± 0.007 c	1.01 ± 0.007 c	4.98 ± 2.87 c	103.77 ± 14.42 a	22.07 ± 10.22 c

[Z] Means within each column followed by the same letter are not significantly different according to All Pairs, Tukey Honestly Significant Difference at 0.05 confidence level.

Less than 20% of CMBS population reached adult stage on 'Tuscarora', 'Natchez' seedlings, and 'Fantasy' seedlings (Figs 1, 2 and 3). The majority of pupae and male adults observed on 'Natchez' seedlings were from day 20 to day 60 (Fig. 2), while all the pupae and adults were observed after day 50 on 'Fantasy' seedlings (Fig. 3).

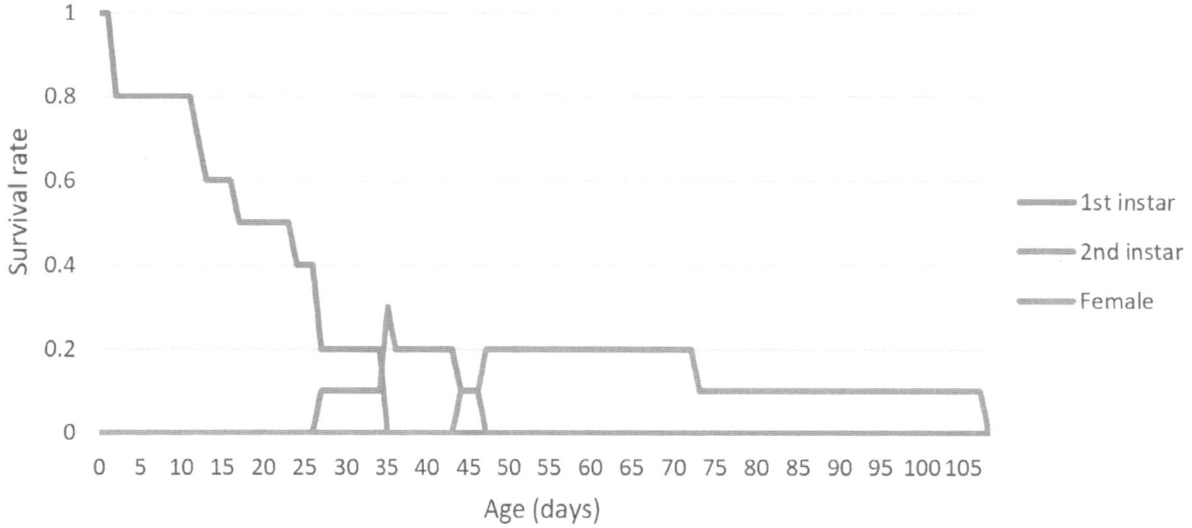

Figure 1. Age-stage specific survival rate (*Sxj*) of *Acanthococcus lagerstroemiae* reared on *Lagerstroemia* 'Tuscarora' at laboratory conditions of 25° C and 250 µmol m^{-2} s^{-1} light with a photoperiod of 12:12 (light: dark).

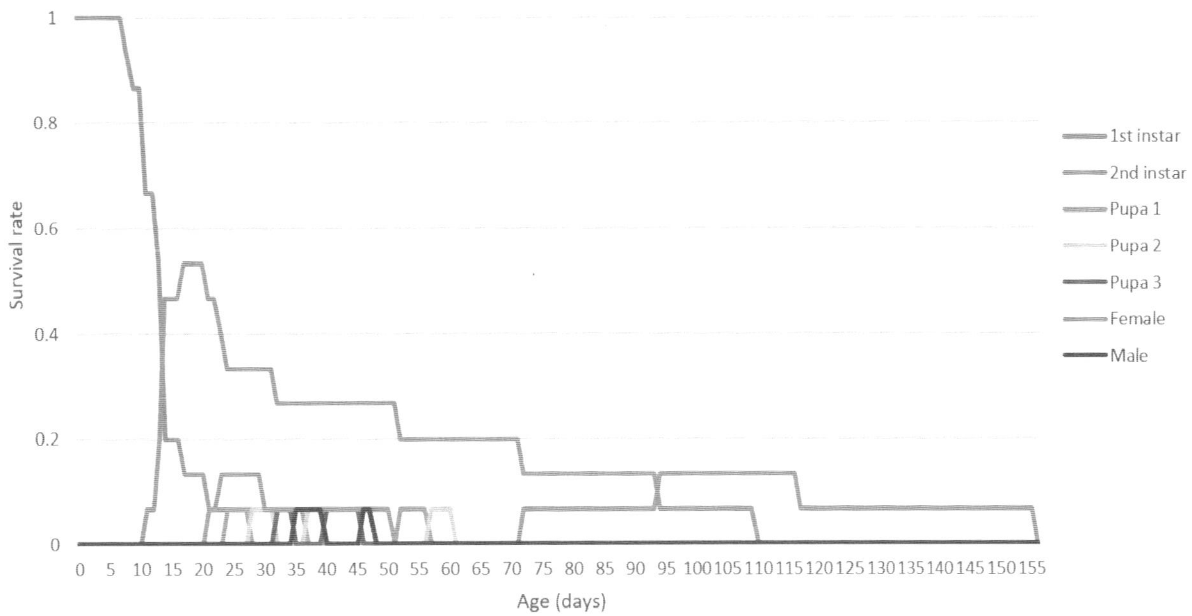

Figure 2. Age-stage specific survival rate (*Sxj*) of *Acanthococcus lagerstroemiae* reared on *Lagerstroemia* 'Natchez' seedlings at laboratory conditions of 25° C and 250 µmol m^{-2} s^{-1} light with a photoperiod of 12:12 (light: dark).

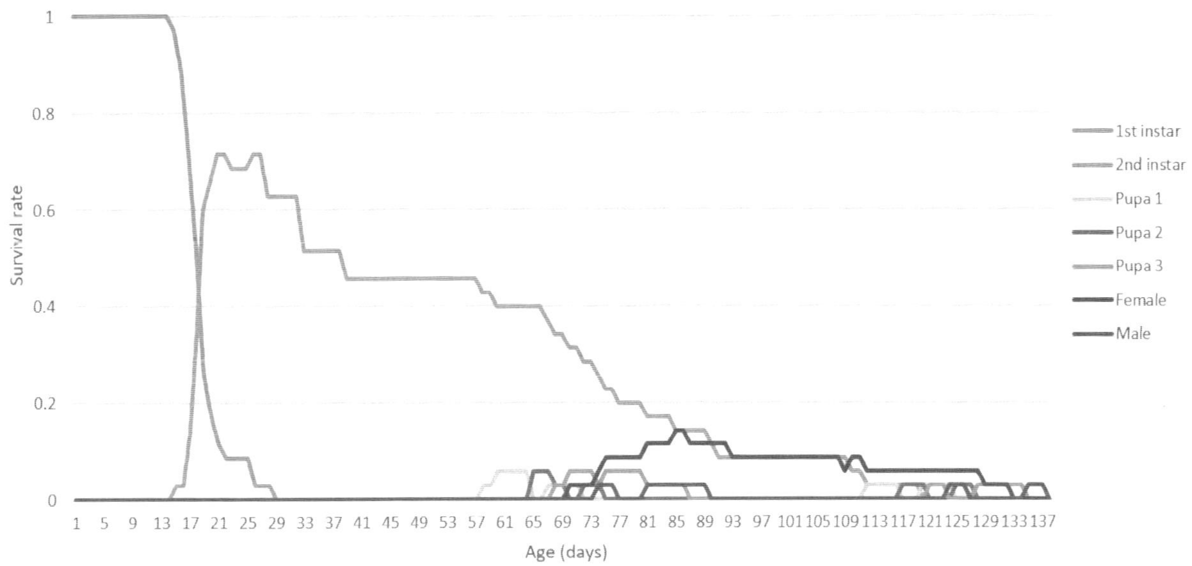

Figure 3. Age-stage specific survival rate (*Sxj*) of *Acanthococcus lagerstroemiae* reared on *Lagerstroemia* 'Fantasy' seedlings at laboratory conditions of 25° C and 250 µmol m^{-2} s^{-1} light with a photoperiod of 12:12 (light: dark).

Net reproduction rate (R_o), the intrinsic rate of increase (r), and the finite rate of natural increase (λ) are population parameters used for projecting the reproductive potential of insects. R_o represents the number of offspring that an individual (including male and female) within the population could produce over its lifetime, while r and λ describe the population growth rate as time approaches infinity and population reaches stable status. Crapemyrtle bark scale with highest population growth rate was found on ‘Natchez’ seedlings, while the lowest was on ‘Fantasy’ seedlings (Table 3).

Overall, the males had shorter life spans compared to the females, which suggested the specific role of adult male in the sexual reproduction of CMBS. The longest-lived females observed were 128, 156, and 143 days on ‘Tuscarora, ‘Natchez’ seedlings, and ‘Fantasy’ seedlings, respectively. The rearing experiments were proven successful in supporting the development of CMBS as the insect was able to complete its life cycle under the experimental conditions in this study.

In this study, the detailed life history data of CMBS on different *Lagerstroemia* hosts were collected and subjected to the age-stage, two-sex life table analysis, in order to provide a comprehensive understanding of CMBS population development at laboratory conditions. The ecological data of CMBS obtained in this study could aid field observations to project the insect population dynamics in the field, and to develop effective control and management strategies for CMBS.

Acknowledgement

This work is supported by the Specialty Crop Research Initiative project ‘Systematic Strategies to Manage Crapemyrtle Bark Scale, An Emerging Exotic Pest’ [grant no. 2017-51181-26831/project accession no. 1013059] from the U.S. Department of Agriculture (USDA) National Institute of Food and Agriculture. Any opinions, findings, conclusions, or recommendations expressed in this publication are those of the author(s) and do not necessarily reflect the view of the U.S. Department of Agriculture.

Literature Cited

Al-Snafi, A. E. (2019). A review on Lagerstroemia indica: A potential medicinal plant. *IOSR* J. Pharmacy *9*:36-42.

Borchsenius, N. S. (1960b). Contribution to the coccid fauna KNR [Chinese Peoples' Republic]. 4. Hard and soft scales, harmful fruit and grape culture in northeast and east KNR. Acta Entomol. Sinica *10*:214-218.

Cabrera, R. I. (2002). Evaluating and promoting the cosmopolitan and multipurpose *Lagerstroemia*. Paper presented at the XXVI International Horticultural Congress: Nursery Crops; Development, Evaluation, Production and Use 630.

Chi, H., & Liu, H. J. B. I. Z. A. S. (1985). Two new methods for the study of insect population ecology. Bull. Inst. Zool., Academia Sinica *24*: 225-240.

Chi, H., & Su, H.-Y. J. E. e. (2006). Age-stage, two-sex life tables of Aphidius gifuensis (Ashmead) (Hymenoptera: Braconidae) and its host *Myzus persicae* (Sulzer) (Homoptera: Aphididae) with mathematical proof of the relationship between female fecundity and the net reproductive rate. Populat. Ecol. *35*: 10-21.

Chi, H. J.. (2020). TWOSEX-MSChart: a computer program for the age-stage, two-sex life table analysis.

Chi, H. (1988). Life-table analysis incorporating both sexes and variable development rates among individuals. Environ. Entomol. *17*: 26-34.

EDDMapS. (2019). Early detection and distribution mapping system.

Harcourt, D. J. (1969). The development and use of life tables in the study of natural insect populations. Ann. Rev. Entomol. *14*:175-196.

Jiang, N., and Xu, H. (1998). Observertion on *Eriococcus lagerostroemiae* Kuwana. J. Anhui Agricult. College, *25*: 142-144.

Jiao, H., and Zhang, Y. (2011). Occurrence and control strategy of crapemyrtle bark scale in Baoji area. Shaanxi J. Agric. Sci., *57*:107-108.

Knox, G. W. (2000). Crapemyrtle in Florida: University of Florida Cooperative Extension Service, ENH-52. https://edis.ifas.ufl.edu/mg266

Kuwana, S. I. (1907). Coccidae of Japan, I. A synoptical list of Coccidae of Japan with descriptions of thirteen new species. Bull. Imperial Cen. Agricult. Exp. Stn. Japan. *1*:177-212.

Pooler, M. (2007). Crapemyrtle. In: Flower Breeding and Genetics, ed. Anderson, N.O., Springer Inc. pp. 439-457 https://www.springer.com/gp/book/9781402044274

Wade, G. L., and Williams-Woodward, J. (2009). Crape myrtle culture. University of Georgia Extension, Cir. 944. https://extension.uga.edu/publications/detail.html?number=C944&title=Crape%20Myrtle%20Culture

Wang, X., Dean, D., Wadl, P., Hadziabdic, D., Scheffler, B., Rinehart, T.,and Trigiano, R. J. H. (2010). Development of microsatellite markers from crape myrtle (*Lagerstroemia* L.). J. Amer. Soc. Hort. Sci. *45*:842-844.

Wang X., Wadl, P.A, Pounders, C., Trigiano, R.N., Cabrera, R. I., Scheffler, B.E. Pooler, M., and Rinehart, T. A. (2011). Evaluation of genetic diversity and pedigree within crapemyrtle cultivars using simple sequence repeat (SSR) markers. J. Amer. Soc. Hort. Sci. *136*:116-128.

Wang, Z., Chen, Y., Gu, M., Vafaie, E., Merchant, M., and Diaz, R. (2016). Crapemyrtle Bark Scale: A New threat for crapemyrtles, a popular landscape plant in the U.S. Insects. 7(4), 78. doi:10.3390/insects7040078

Wu, B., Xie, R., Knox, G. W., Qin, H., and Gu, M. Crapemyrtle bark scale (Hemiptera: Eriococcidae) showed different acceptance among crapemyrtle species significantly. status (unpublished; manuscript in preparation).

Xie, R., Wu, B., Dou, H., Liu, C., Knox, G. W., Qin, H., & Gu, M. J. I. (2020). Feeding Preference of Crapemyrtle Bark Scale (*Acanthococcus lagerstroemiae*) on different species. Insects 11(7), 399. doi: 10.3390/insects11070399

Zhang, Z., & Shi, Y. (1986). Studies on the morphology and biology of *Eriococcus lagerstoemiaea* Kuwana . J. Shandong Agricul. Univer, 2.

Combined Proceedings IPPS 70:81-86. 2020.

Effects of Gibberellic Acid and Cold Stratification on Sparkleberry (*Vaccinium arboreum*) Germination Under Different Collection Times

Ping Yu[1], Lin Li[2], Qiansheng Li[3] and Mengmeng Gu[3]

[1]Department of Horticulture Sciences, Texas A&M University, College Station, TX, 77843, USA; [2] College of Architectural Arts, Guangxi Arts University, Nanning, Guangxi 530007, China; [3]Department of Horticulture Sciences, Texas A&M AgriLife Extension Services, College Station, TX, 77843, USA

yuping520@tamu.edu

Keywords: Blueberry rootstock, seed germination

Summary

Sparkleberry has the potential to be used as commercial blueberry's (*Vaccinium spp.*) rootstock due to its wider adaptation to the environment, tolerance to higher pH, and its singular architecture, which can reduce blueberry yield loss during mechanical harvesting. There is little information in the literature on seed germination of sparkleberry. We report that gibberellic acid and cold stratification work synergistically to increase sparkleberry germination for seed collected in Texas during November and December. The optimal germination treatments for sparkleberry seeds collected from November was 500 mg L^{-1} gibberellic acid (GA_3) followed by cold stratification for 9 weeks, which had a 70.4% emergence percentage.

IPPS Vol. 70 – 2020

INTRODUCTION

Sparkleberry (*Vaccinium arboretum*) is a shrub in *Vaccinium* family native from climatic zone seven to nine in the United States. Sparkleberry can be found from southern Virginia to southeastern Nebraska, from Florida to eastern Texas (Dirr, 1990). Sparkleberry has the potential to be used as commercial blueberry's (*Vaccinium spp.*) rootstock due to its wider adaptation to the environment (Griffin and Blazich, 2008; Casamali et al., 2016a). Sparkleberry has a broader tolerance to soil pH and requires lower soil organic matter than blueberry. Also, unlike branchy blueberry, sparkleberry has a singular architecture, which can reduce blueberry yield loss during mechanical harvesting (Brooks and Lyrene, 1998; Yang et al., 2014; Casamali et al., 2016b).

How can we propagate sparkleberry to fulfill its potential as blueberry rootstock? It is hard to root sparkleberry either via softwood cuttings or hardwood cuttings. Lyrene (1998) reported difficulty in propagating sparkleberry via softwood cuttings. In our previous trials, none of sparkleberry hardwood cuttings or softwood cuttings from plants in the native habitat rooted.

Little information about sparkleberry seed germination can be found in the literature. The only study on sparkleberry seed germination in recent years reported that the highest seed germination rate was 43%, which was obtained by cold stratification (the process of subjecting seeds to both cold and moist conditions) for 90 days (Yang et al., 2014). Treating *Vaccinium* species with gibberellic acid (GA_3) could increase the germination of *Vaccinium* species in some cases (Griffin and Blazich, 2008).

Therefore, this study was to evaluate the effects of GA_3 and cold stratification on germination of sparkleberry seeds collected during different times. The goal of this study was to establish guidelines and provide valuable information for sparkleberry mass production in nurseries for blueberry rootstock.

MATERIALS AND METHODS

Plant Material

Sparkleberry seeds were collected from ripe berries on plants grown in natural areas of the Woodland Hills Park (College Station, TX, USA) in November and December, 2018. Viable seeds were obtained by washing seeds several times and selecting seeds at the bottom of water as Ciordia et al. (2006) found that floating seeds had much-reduced viability. Seeds were surface sterilized with 5% regular Clorox bleach for 10 minutes, then rinsed with deionized water (DI water) and dried in open air at room temperature for two days.

Experimental Treatment

After being dried, for each treatment, 50 seeds were put into 10 cm petri dish filled with 20 mL of (GA_3) solutions at 500 and 1,000 mg L^{-1} for 24 h, and the same amount of DI water was used as GA_3 0 mg L^{-1} treatment (Fig. 1). Seeds were then rinsed with DI water and sown in 4-in. pots (dimensions: top 7.5cm, bottom 6cm, depth 8.2cm with four holes underneath) filled with commercial germination mix (Pro-mix HP with Mycorrhizae, BWI Companies, USA) on April 10, 2019. After adding a thin layer of media on the top of the seeds, pots with GA_3 treated seeds were put either into the greenhouse (0 week of cold stratification) or the cold storage for cold stratification for 3, 6 and 9 weeks. After cold stratification, pots were then taken out and placed in the greenhouse. Pots were watered with DI water as needed throughout the experiment.

Figure 1. Sparkleberry (*Vaccinium arboreum*) seed collected in November and December in 2018, subjected to different gibberellic acid (GA_3) (0, 500, and 1,000 mg L^{-1}) in petri dishes (A); Sparkleberry (*Vaccinium arboreum*) seed collected in November and December in 2018, germinated after 3 weeks of cold stratification with GA_3 treated at 0, 500, and 1,000 mg L^{-1} (B).

Experimental Design

The experiment was arranged in a two-factor split-plot design with the GA_3 as the main plot and cold stratification as the subplot with five replicates for each treatment. The two factors were GA_3 concentration (0, 500, and 1,000 mg L^{-1}) and cold stratification (0, 3, 6, and 9 weeks). Each treatment in the experiment contained 50 seeds. Seeds collected in November and December were analyzed separately.

Data Collection

Data was collected by counting the numbers of germinated seeds (cotyledon emerged from the media) twice a week starting at week 0 for 0, 3, 6, and 9 weeks of cold stratification until week five when no more seeds came out. The emergence percentage (EP) was calculated by the following formula: EP = (No. of emerged seeds/total No. of seeds) × 100% and the emergence index (EI) were calculated as following: EI $=\sum_{i=1}^{n}(EPi/Ti)$, where EP_i is EP on day i ($i \geq 2$), and Ti is the number of days after sowing.

Data Analysis

Under two seed collection times (November and December), analysis of variance (ANOVA) was used to test the significance of treatments and interaction effects on germination. When the ANOVA showed difference, the multiple honest significant difference (HSD) Tukey's test was applied with the calculation of significant differences among means at $p \leq 0.05$.

RESULTS

There were two ways interactions (GA_3 and stratification) in the emergence percentage (EP) and emergence index (EI) for both November and December seeds (Table 1). GA_3 and stratification had significant impacts on EP and EI for both November and December seeds.

Table 1: A summary of the statistical significance of treatment factors on emergence percentage (EP, %), emergence index (EI, %), and number of days needed for germination. *, **, *** indicated significant difference according to Tukey HSD multiple comparison test at $p \leq 0.05$, ≤ 0.01, and ≤ 0.001, respectively.

	November		**December**	
Source	Emergence (%)	Emergence index	Emergence (%)	Emergence index
GA_3	***	***	*	**
Stratification	***	**	***	***
GA_3 × Stratification	***	***	***	***

For seed collected in November, cold stratification significantly increased seed EP (Fig. 2 A, B, C, D). For seed had been stratified for 0 week, 3 weeks, and 9 weeks, GA_3 concentration had significant influence on EP (Fig. 2 A, B, D). For seed had been stratified for 0 week and 3 weeks, 1,000 mg L^{-1} GA_3 had significantly higher EP than those without GA_3 (Fig. 2 A, B).

Seed treated with GA_3 at 1,000 mg L^{-1} for 3 weeks could reach an EP at 58.8%. There was no significant difference for 6-week treated seed among 0, 500 or 1,000 mg L^{-1} GA_3 treatments (Fig. 2 C). Seed treated with GA_3 at 500 mg L^{-1} for 6 weeks could reach an EP at 54.4%. For 9 weeks seed, 500 mg L^{-1} GA_3 had significantly higher EP than those with GA_3 at 0 or 1,000 mg L^{-1} (Fig. 2 D). Seed treated with GA_3 at 500 mg L^{-1} for 9 weeks had the highest EP (70.4%).

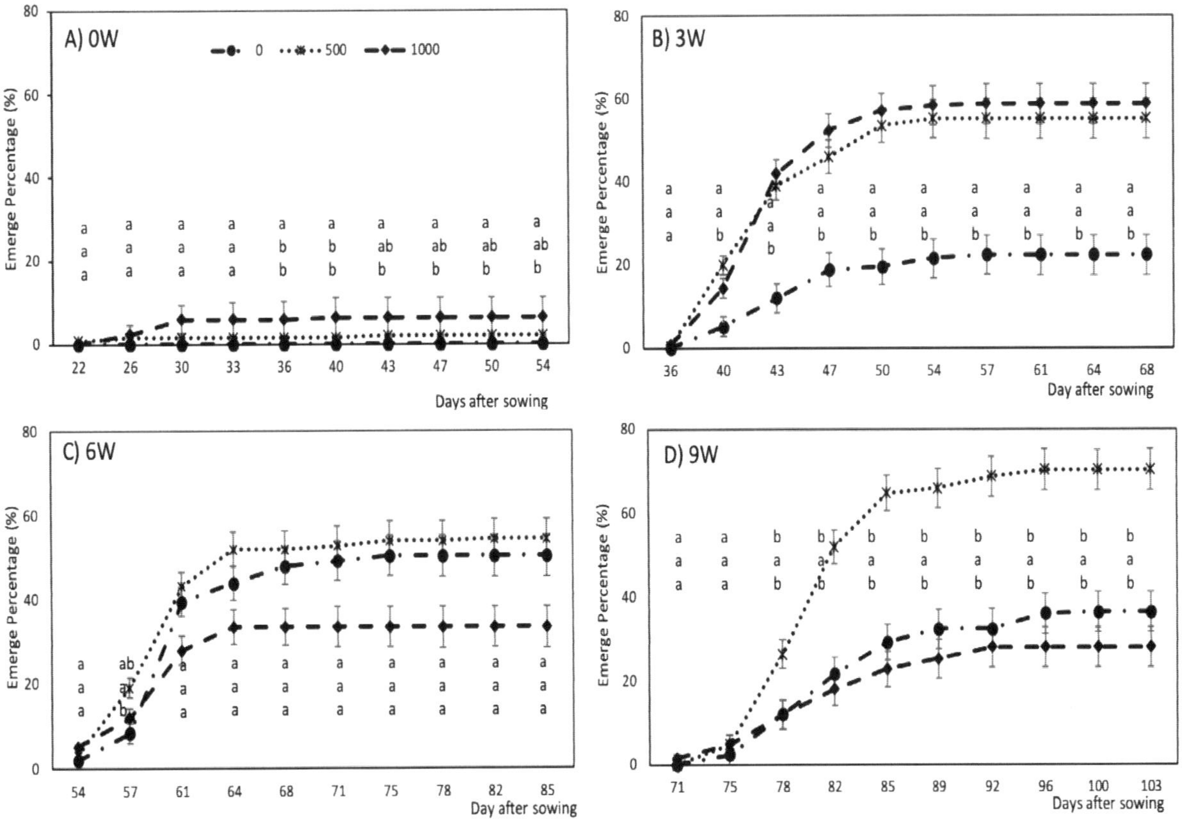

Figure 2. Emergence percentage (± standard error) for sparkleberry (*Vaccinium arboreum*) seed collected in November with cold stratification for 0 (A), 3 (B), 6 (C) and 9 (D) weeks and gibberellic acid (GA_3) at 0, 500, and 1,000 mg L^{-1} (indicated by solid line, dash line, and dash line with diamond, respectively). The same letters on the same day (in an order of 0, 500, 1,000 mg L^{-1} from the bottom to the top) indicate no significant difference among GA_3 treatments (0, 500, 1,000 mg L^{-1}) according to Tukey HSD multiple comparison test at $p \leq 0.05$.

CONCLUSION

In conclusion, we found that gibberellic acid and cold stratification could work synergistically to increase sparkleberry germination for seed collected in both November and December (Data not shown).

The optimal germination treatments for sparkleberry seed collected from November was 500 mg L^{-1} GA_3 followed by stratification for 9 weeks, which had a 70.4% emergence percentage.

Literature Cited

Brooks, S.J. and P.M. Lyrene. (1998). Derivatives of *Vaccinium arboreum*× *Vaccinium* section cyanococcus: I. Morphological characteristics. Journal of the American Society for Horticultural Science *123*:273-277.

Casamali, B., J.G. Williamson, A.P. Kovaleski, S.A. Sargent, and R.L. Darnel. (2016). Mechanical harvesting and postharvest storage of two southern highbush blueberry cultivars grafted onto *Vaccinium arboreum* rootstocks. HortScience *51*:1503-1510.

Ciordia, M., M.d. Lucas, V. Mateos, L. Rodriguez, J. Garcia, and J. Majada. (2006). Optimization of germination requirement and seed production of wild-type *Vaccinium myrtillus*. Acta horticulturae *715*:85-88.

Dirr, M.A., (1990). Manual of woody landscape plants: their identification, ornamental characteristics, culture, propagation and uses. Stripes Publishing, Champaingn, 1187 p.

Griffin, J.J. and F.A. Blazich. (2008). *Vaccinium* L. Woody plant seed manual. *US Department Agricultural and Forest Service*. Washington. EE. UU:1154-1159.

Lyrene, P.M. (1998). Germination and growth of sparkleberry seedlings (*Vaccinium arboreum* Marsh). Fruit Varieties Journal *52*:171-178.

Yang, W., H. Andrews, and A. Basey. (2014). Blueberry rootstock: Selection, evaluation, and field performance of grafted blueberry plants. Acta horticulturae *1117*:119-123.

Combined Proceedings IPPS 70:87-94. 2020.

Lessons from Nature: Studies on Mangrove Trees and Biotechnology

Prakash P. Kumar and Pannaga Krishnamurthy

Department of Biological Sciences, and NUS Environmental Research Institute, National University of Singapore, 10 Science Drive 4, Singapore 117543

prakash.kumar@nus.edu.sg

Keywords: *Avicennia*, *CYTOCHROME P450,* salinity, roots, ecology suberin, salt glands

Summary

Salinity is an abiotic stress that reduces the growth and productivity of crop plants worldwide. Mangrove trees such as *Avicennia officinalis* exhibit remarkable ability to grow in saline environment by means of various adaptations. Such adaptations include secretion of excess salt from the leaf surface via specialized salt glands, as well as the ability to exclude ~95% salt from seawater due to enhanced hydrophobic root barriers (suberin lamellae) in their roots. Certain cytochrome P450 enzymes play a key role in biosynthesis of suberin precursors. Knowledge gained by studying mangroves can be used for biotechnological applications. Thus, we identified several *CYTOCHROME P450* (*CYP*) genes that were induced by salt treatment in *A. officinalis* roots. Using appropriate *Arabidopsis* mutants, such as *atcyp94b1,* we characterized the function of *CYP94B1* gene in regulating suberin biosynthesis. The *atcyp94b1* mutant seedlings showed salt sensitivity with reduction in root elongation. When treated with salt, their roots exhibited reduced suberin lamellae and Casparian bands. Heterologous expression of the coding sequence of *A. officinalis CYP94B1* in *atcyp94b1* resulted in rescuing of the salt sensitive phenotypes, indicating the involvement of CYP94B1 enzyme in suberin biosynthesis. Additionally, we have expressed selected genes from the mangrove in rice, and the transgenic rice plants acquired higher salinity tolerance. These findings opened up additional strategies for salinity tolerant crop plants in the future. Our current efforts in these projects will be discussed.

IPPS Vol. 70 – 2020

INTRODUCTION

This is the text version of an invited talk delivered online at the IPPS Western Region, USA, 27-29 October 2020. I thank the organizers for giving me an opportunity to share some of our research findings at this forum. Our major research interests can be summarized under three categories: 1) Plant hormone signal transduction and control of plant development, with the focus on cytokinin and gibberellin signaling (Ravindran *et al.*, 2017). 2) Mechanism of salt tolerance in the mangrove tree species, namely, *Avicennia officinalis*. This project includes the mechanism of ultrafiltration of salt at the roots and salt secretion at the leaf via salt glands (Tan *et al.*, 2013). 3) The findings from the above studies are being applied for devising crop improvement strategies (biotechnological and molecular) using rice as the crop model. Also, other findings such as using molecular biology knowledge such as water and ion transporters from the mangrove for recombinant protein production and preparing biomimetic membranes incorporating aquaporin protein for water purification in the long-term. This talk will cover some of our findings from the second and third research areas of interest.

According to an estimate ~2,000 hectares/day of irrigated land is degraded by increased soil salinity across 75 countries. Every week, an area larger than Manhattan is lost due to salinity. Thus, currently an estimated ~62 million hectares (which corresponds to 20%) of the world's irrigated lands are affected by high salinity problems (up from 45 million hectares in 1990s). The corresponding lost crop value/year is $27 billion annually. According to the United Nations University Institute for Water, Environment and Health, "We can't afford not to restore the productivity of salt-affected lands". It is well established that salinity stress adversely affects plants in multiple fronts, e.g., genetic, physiological and morphological status of crop plants, ultimately leading to major losses in crop yield. Therefore, it is important to develop stress-tolerant varieties to ensure future food security.

One of the strategies to develop stress tolerance in plants is pathway engineering, which may be accomplished by biotechnology or by molecular breeding techniques. Specific plant hormone signaling pathways are important to alter the survival, growth and yield of plants. Cells perceive the stress-signals from the soil and environment and various endogenous factors then react in order to attempt to suitably modify the growth and physiology of plants (Munns and Tester, 2008). For example, salinity and drought stresses upset ion balance in the cells. One will need to use as much data as possible to attempt to modify the multitude of endogenous factors to remediate the stress. Among the possible strategies an important one is to try and modify genes that affect tolerance, e.g., using genomic breeding approaches. Thus, the use of 'big data' can help to derive strategies for breeding crops having higher tolerance to stresses in the years to come. Where will we be able to obtain such big data? Various omics data sets (e.g., proteomics, genomics, transcriptomics, metabolomics) are becoming available for crop plants. Such information from naturally salt tolerant plants (e.g., mangroves) can be useful to confer stress tolerance to crop plants. As a brief background for the mangrove species, they naturally grow right in the seawater. They are able to thrive in such high salinity that would not permit growth of crop plants, because mangroves possess multiple adaptations, including propagules that germinate while still attached to the parent plants so that when they are shed, they can readily establish themselves as seedling on the muddy floor, or will be carried by ocean currents to long distances before establishing themselves on distant shores. Seedlings grown from propagules in the greenhouse are shown in Figure 1.

Figure 1. Greenhouse-grown seedlings of *Avicennia officinalis*. Propagules collected from the trees were sown in the soil and allowed to grow for about 2 months.

Mangroves are highly recalcitrant species and thus far they have not been successfully propagated by tissue culture. Natural propagation occurs by means of fruits/propagules, which are usually viviparous (germinate on the mother plant as mentioned above). One of our earlier studies had revealed that several species of mangroves have propagules that can float and they are transported to long distances in sea currents.

Desalination in mangroves via ultra-filtration at the roots

Plants inhabiting a saline environment literally lead a 'Life in a pickle'! They are under constant stress of high solute (salt) soil environment and have to deal with a resultant water-deficit condition. This limits the growth and survival of most plants, but mangroves (that are categorized as halophytes) thrive in marine and estuarine shorelines. They accomplish this because of their special adaptations, including, 1) Their roots carry out ultrafiltration of salt (prevent salt uptake); and 2) some mangroves can secrete excess salts via specialized salt glands on leaves (Figure 2). These represent a form of natural desalination process at work! Salt glands are microscopic structures located mainly on leaf surfaces. And, the roots of mangroves are able to carry out ultrafiltration of salt. Therefore, we are interested in understanding how salt glands and roots work. These insights can help in developing stress tolerant plants in the long-term.

Figure 2. Two-month-old *Avicennia officinalis* seedlings were transferred to pots with sand and allowed to adapt for two days before treating with 500 mM NaCl. Salt secretion could be visualized as white salt crystals on the leaves after two weeks of salt treatment.

Ultrafiltration at the mangrove roots represents the first physiological defense for plants growing in saline soil. This refers to exclusion of salt prior to uptake of water into the xylem. Our experimental results have shown that *Avicennia* roots are able to exclude ~95% of the salt from seawater because of the presence of physical barriers, namely, suberin lamellae in specific root cell layers (e.g., endodermis, exodermis). We have published several papers reporting that increased suberization in response to salt treatment helps in efficient root filtration of salt, and that a series of genes are differentially expressed in response to salt (Krishnamurthy *et al.*, 2014a; Krishnamurthy *et al.*, 2017; Krishnamurthy *et al.*, 2014b).

Unravelling the mechanism underlying salt uptake at the roots of mangroves

Control of growth needs plant water relations to be properly regulated. Ion transporters are involved in regulating water relations as well as uptake of nutrients (e.g., K^+, NO_3^-). Other genes that were identified include those that encode membrane proteins, such as aquaporins and ion channels involved in salt uptake and secretion. We are studying several genes in this category, and some of these genes were used in our biotechnological efforts for generating salt tolerant plants

(Krishnamurthy et al., 2019; Rajappa *et al.*, 2020). However, due to time constraints, these findings will not be discussed further in this talk.

Regulation of apoplastic barrier formation and identification of the molecular mechanism

We will discuss some details from our attempts at understanding the mechanism of salt ultrafiltration. We discovered several genes encoding for Cytochrome P450 monooxygenases as being specifically induced by salt based on a differential gene expression analysis using transcriptomics, followed by gene function analysis.

When we subjected our RNAseq data (transcriptomics) to functional analysis using the KEGG pathway, we discovered that several genes for suberin biosynthesis are differentially regulated by salt treatment in the mangrove roots. It was previously known that enzymes, such as, CYP94A1, CYP86B1 are required for biosynthesis of suberin precursors.

We carried out most of the molecular biological and transgenic plant studies using *Arabidopsis thaliana* as the experimental species. This is due to the fact that: 1) homologs of most of these genes can be identified in *Arabidopsis*, 2) relevant mutants in the genes of interest are available in this species, and 3) mangrove trees cannot be subjected to transgenic plant production and genetic analyses. Therefore, other than simple gene expression analysis in *Avicennia*, the bulk of the results are from *Arabidopsis* studies. The key findings are presented in summary form in the sections below. Details have just been published (Krishnamurthy et al., 2020) and readers may wish to refer to this publication.

Our results showed that *CYP94B1* gene is consistently upregulated by salt treatment, especially in the roots of *Avicennia* and *Arabidopsis*. The enzyme encoded by this gene, namely, CYP94B1 catalyzes synthesis of omega-hydroxy fatty acids, which are suberin monomers. Detailed gene expression analysis showed that *AtCYP94B1* is preferentially expressed in the endodermis under salt treatment. We obtained the relevant mutant *atcyp94b1* from the stock center. This mutant exhibits salt sensitive phenotype, as illustrated by inhibition of seedling root elongation when grown on nutrient medium with 75 mM NaCl. This phenotype could be rescued by genetic complementation accomplished by trans-genically expressing either *AoCYP94B1* or *AtCYP94B1* gene. Transgenic lines expressing mangrove CYP exhibit better growth under NaCl conditions.

Because of the involvement of the gene in regulating suberin biosynthesis, we examined suberin content in the various seedlings using confocal laser scanning microscopy. Suberin deposition was significantly reduced in the mutants. But, this was restored to normal levels when the *CYP* gene was expressed in the roots of *atcyp94b1* mutant. We showed that the reduced suberin in the roots of *cyp94b1* causes more uptake of salt as indicated by the loading of the tracer apoplastic dye (FDA) to pass through the endodermis and enter the pericycle cells. The transgenic lines showed that along with the restoration of suberin in the endodermis cells, the uptake of the tracer (indicative of salt up-take) was also significantly reduced. This is the ultrafiltration mechanism that confers enhanced salt tolerance to the plants.

Can such salt tolerance be transferred to crop plants such as rice?

We were interested to test if a similar salt tolerance mechanism can be conferred to rice seedlings by introducing the *CYP* genes. Accordingly, we generated transgenic rice plants expressing *AoCYP94B1* gene and our results showed that these plants have increased salt tolerance as evidenced by good growth under 100 mM NaCl treatment. We also showed that such salt treatment to normal wild type plants leads to stunted growth of seedlings. The transgenic rice seedlings exhibited better seedling height and better recovery growth after 3 weeks of salt treatment.

One-month-old *pUbi::AoCYP94B1* rice seedlings were treated with salt for 21 days and then allowed to recover by removing the salt in the water (Figure 3 and Table 1).

After salt treatment **Resuming growth**

Figure 3. Four-week-old hydroponically grown wild-type and transgenic *pUBI::AoCYP94B1* rice plants after 21 days of 100 mM NaCl treatment and an additional 10 days of recovery growth without NaCl.

Table 1. Survival rates of wild-type and transgenic rice plants (from Fig 3) after salt treatment and recovery.

Genotype	Survival rate (%) ± SD
WT	36.7 ± 6.3
pUBI::AoCYP94B1	74.9 ± 11.6

More importantly, we were able to show that the transgenic rice roots expressing the mangrove *CYP* gene (*AoCYP94B1*) have increased suberin deposition in both endodermis and exodermis. Therefore, the higher salinity tolerance of these transgenic plants correlates with increased suberin deposition in their roots, similar to our observations in *Arabidopsis*. These results collectively show that salt tolerance trait can be conferred to rice and other plant species by introducing *CYP94B1*.

What is the molecular regulatory factor of AtCYP94B1 gene in Arabidopsis?

Lastly, we attempted to determine how *AtCYP94B1* gene is regulated in *Arabidopsis*. A bioinformatic analysis showed that *AtCYP94B1* promoter contains several stress related transcription factor (TF) binding sites, such as WRKY. We found that WRKY33 is involved in regulating the gene expression. Coincidentally, salt induces *WRKY33* gene similar to the induction of *CYP94B1* gene. Also, we obtained data to show that *AtCYP94B1* expression is suppressed in *atwrky33* mutants, and that AtWRKY33 directly binds to *AtCYP94B1* gene promoter and induces its expression (Y1H, ChIP, Luciferase assays) confirming this molecular regulation process.

CONCLUSION

- Ultrafiltration in the mangrove *Avicennia* roots helps to exclude ~95% salt from seawater.
- Endodermis and exodermis show increased suberization in response to salt treatment, which help in ultrafiltration of salt at the roots of the mangrove tree.
- Several *Cytochrome P450* (regulating biosynthesis of ω-hydroxylases involved in suberin biosynthesis) and *WRKY* genes are upregulated by salt (transcriptomics data).
- *AoCYP94B1* (also *AtCYP94B1*) rescues salt sensitive and reduced suberin phenotypes of the *Arabidopsis atcyp94b1* mutant roots.
- WRKY33 transcription factor that is co-induced by salt, helps to regulate *AtCYP94B1* gene.
- Transgenic rice plants expressing mangrove At*CYP94B1* exhibit increased suberin deposition and enhanced salinity tolerance.
- Understanding the physiological and molecular mechanisms of stress tolerance in mangroves (that have special adaptations to survive in the saline environment) may help to generate future crop plant varieties with higher abiotic stress tolerance.
- We need to combine gene technologies with novel propagation and crop management techniques to enable plants to grow well with reduced inputs and tolerate serious abiotic stresses.

Funding acknowledgement: This research grant is supported by the Singapore National Research Foundation under its Environment & Water Research Programme and administered by PUB, Singapore's National Water Agency, NRF-EWI-IRIS (R-706-000-010-272 and R-706-000-040-279).

Literature Cited

Krishnamurthy, P., Jyothi-Prakash, P.A., Qin, L., He, J., Lin, Q., Loh, C.S., and Kumar, P.P. (2014a). Role of root hydrophobic barriers in salt exclusion of a mangrove plant *Avicennia officinalis*. Plant, Cell and Environment *37*, 1656-1671.

Krishnamurthy, P., Tan, X.F., Lim, T.K., Lim, T.M., Kumar, P.P., Loh, C.S., and Lin, Q. (2014b). Proteomic analysis of plasma membrane and tonoplast from the leaves of mangrove plant *Avicennia officinalis*. Proteomics *14*, 2545-2557.

Krishnamurthy, P., Mohanty, B., Wijaya, E., Lee, D.Y., Lim, T.M., Lin, Q., Xu, J., Loh, C.S., and Kumar, P.P. (2017). Transcriptomics analysis of salt stress tolerance in the roots of the mangrove *Avicennia officinalis*. Scientific Reports *7*, 10031.

Krishnamurthy, P., Vishal, B., Ho, W.J., Lok, F.C.J., Lee, F.S.M., and Kumar, P.P. (2020). Regulation of a cytochrome P450 gene *CYP94B1* by WRKY33 transcription factor controls apoplastic barrier formation in roots to confer salt tolerance. Plant Physiology *184*, 2199-2215.

Krishnamurthy, P., Vishal, B., Khoo, K., Rajappa, S., Loh, C.S., and Kumar, P.P. (2019). Expression of *AoNHX1* increases salt tolerance of rice and Arabidopsis, and bHLH transcription factors regulate *AtNHX1* and *AtNHX6* in Arabidopsis. Plant Cell Reports *38*, 1299-1315.

Munns, R., and Tester, M. (2008). Mechanisms of salinity tolerance. Annual Review of Plant Biology *59*, 651-681.

Rajappa, S., Krishnamurthy, P., and Kumar, P.P. (2020). Regulation of *AtKUP2* expression by bHLH and WRKY transcription factors helps to confer increased salt tolerance to *Arabidopsis thaliana* plants. Frontiers in Plant Science *11*, 1311.

Ravindran, P., Verma, V., Stamm, P., and Kumar, P.P. (2017). A novel RGL2-DOF6 complex contributes to primary seed dormancy in *Arabidopsis thaliana* by regulating a GATA transcription factor. Molecular Plant *10*, 1307-1320.

Tan, W.K., Lin, Q., Lim, T.M., Kumar, P., and Loh, C.S. (2013). Dynamic secretion changes in the salt glands of the mangrove tree species *Avicennia officinalis* in response to a changing saline environment. Plant, Cell and Environment *36*, 1410-1422.